Springer Undergraduate Texts in Mathematics and Technology

More information about this series at http://www.springer.com/series/7438

David R. Finston • Patrick J. Morandi

Abstract Algebra

Structure and Application

Springer

David R. Finston
Department of Mathematics
Brooklyn College of the City
University of New York
Brooklyn, NY, USA

CUNY Graduate Center
New York, NY, USA

Patrick J. Morandi
Department of Mathematical Sciences
New Mexico State University
Las Cruces, NM, USA

Additional material to this book can be downloaded from http://extras.springer.com

ISSN 1867-5506 ISSN 1867-5514 (electronic)
ISBN 978-3-319-04497-2 ISBN 978-3-319-04498-9 (eBook)
DOI 10.1007/978-3-319-04498-9
Springer Cham Heidelberg New York Dordrecht London

Library of Congress Control Number: 2014942314

Mathematics Subject Classification (2010): 11T71, 12-01, 16-01, 20-01, 51-01

Printed on acid-free paper

Springer is part of Springer Science+Business Media (www.springer.com)

Preface

This book evolved from our experiences over several years teaching abstract algebra to mixed audiences of mathematics majors and majors in secondary mathematics education at New Mexico State University (the course is required for both groups at NMSU as it is in many institutions of higher learning in the USA) along with our outreach work with Las Cruces area middle and high school mathematics students and teachers. These undertakings left us with a dilemma. While sympathetic to the frustrations expressed by pre-service and in-service teachers with the abstract nature of the standard presentations of the subject matter, and the perception of its irrelevance to pre-college teaching, we maintain that a rigorous grounding in the conceptual framework of algebra is absolutely critical to a high school or middle mathematics teacher's success, both in conveying content to their students and in fostering their enthusiasm and self-confidence for future careers in STEM fields and even public policy. The latter is particularly timely given the ubiquitous use of social media and current controversies over corporate and governmental surveillance. Our solution was to develop the structures and basic theorems of modern algebra through applications that have relevance to daily life (e.g., Identification Schemes, Error Correcting Codes, Cryptography, Wallpaper Patterns) and that directly inform topics that arise in high school or middle school mathematics classes (e.g., Number Theory, Symmetry, Ruler and Compass Constructions).

The result is a text intended for a one semester course in modern algebra that can be used in a variety of contexts. For an audience composed primarily of mathematics majors, the material on identification numbers, modular arithmetic, and linear algebra over arbitrary fields can be covered quickly, so that the chapters on codes defined over finite fields, isometries of the real plane, and ruler and compass constructions (and the associated abstract ring, field, and group theory) can be covered in depth. For an Applied Algebra course, with computer science majors in mind, the material on ruler and compass constructions can be given a lighter treatment so that emphasis can be placed on error detection and correction, cryptography, and isometries (important for computer-aided design). For courses designed for secondary mathematics teachers, the chapters on identification numbers, linear codes, ruler and compass constructions, and isometries (at least through the classification of frieze patterns) introduce groups, rings, and fields through accessible applications and provide ample rigor. A course based on these chapters would also serve programs offering a Master's degree in middle school mathematics education or a Master of Arts in Teaching Mathematics.

Numerous exercises are given after appropriate subsections. An exception is in Chap. 6 on ruler and compass constructions, where some steps in proofs are given as exercises within the text. This is done not only because the requisite drawings take up a lot of text space but also, more importantly, because they're fun. Exercises range from routine verifications and computations to more serious applications of the text material and conceptual issues. Proofs of a few propositions are left as exercises because they give opportunities to employ important techniques that have been used earlier and will arise again. Some of the exercises refer to electronic supplementary materials (ESM) in the form of MAPLE worksheets. The worksheets, which give the reader practice with computations in modular arithmetic, RSA encryption and decryption, and error correction for Reed–Solomon codes, are accessible from this book's page at http://link.springer.com.

While the text is self-contained, references to supplementary sources solely for more background or further study are given at the end of each chapter.

Brooklyn, NY, USA David R. Finston
Las Cruces, NM, USA Patrick J. Morandi

Contents

Chapter 1
Identification Numbers and Modular Arithmetic

The first topic we will investigate is the mathematics of identification numbers. Many familiar things are described by a code of digits; zip codes, items in a grocery store, and books, to name three. One feature to all of these codes is the inclusion of an extra numerical digit, called a *check digit*, designed to detect errors in reading the code. When a machine (or a human) reads information, there is always the possibility of the information being read incorrectly. For example, moisture or dirt on the scanner used by a grocery store clerk can prevent an item's code from being read correctly. It would be unacceptable if, because of a scanning error, customers were charged for caviar when they are buying tuna fish. The use of the check digit allows for the detection of some scanning errors. If an error is detected, the item is re-scanned until the correct code is read.

1.1 Examples of Identification Numbers

There are many types of identification numbers in common use today. We will discuss three of them: the United States Postal Service zip code, the Universal Product Code (UPC) used for consumer products, and the International Standard Book Number (ISBN), exploring their design and capability for error detection.

1.1.1 The USPS Zip Code

The United States Postal Service uses a bar code to read zip codes on mail. The following bar code is that for the Mathematical Sciences Department of NMSU, whose zip code is 88003-8001.

The original version of this chapter was revised: The figure was corrected at the end of chapter opening page. The correction to this chapter is available at https://doi.org/10.1007/978-3-319-04498-9_11

Electronic supplementary material The online version of this chapter (doi:10.1007/978-3-319-04498-9_1) contains supplementary material, which is available to authorized users. The supplementary material can also be downloaded from http://extras.springer.com.

D.R. Finston and P.J. Morandi, *Abstract Algebra: Structure and Application*,
Springer Undergraduate Texts in Mathematics and Technology, DOI 10.1007/978-3-319-04498-9_1

The bar code represents a ten digit number. There are 52 lines in the bar code. The first and last lines are just markers. The remaining 50 lines comprise ten groups of five, and each group of five represents a digit. The first nine digits form the nine digit zip code of the addressee. The tenth digit is a check digit. This digit is computed as follows: the digits forming the zip code are added, and the check digit is the smallest nonnegative integer needed to make the sum be divisible by 10. For example, given the zip code 88003-8001 for the Department of Mathematical Sciences at New Mexico State University, the nine digits sum to 28. Therefore, the check digit, which is not among the nine digits of the zip code, must be 2. Thus, the bar code represents the ten digit number 8800380012 which consists of the full nine digit zip code with the check digit appended at the end.

This scheme allows one to determine the check digit for any nine digit zip code. For example, if we only knew the nine digit zip code 88003-8001, the check digit x would be the number between 0 and 9 such that the sum

$$8+8+0+0+3+8+0+0+1+x$$

was evenly divisible by 10. Since this sum is $28+x$, the only choice for x is to be 2.

The purpose of the check digit is to detect errors in reading the code. For example, suppose that the zip code 8800380012 was incorrectly read as 8800880012, by reading the fifth digit as an 8 instead of as a 3. The sum of the digits would then be $8+8+8+8+1+2=35$, which is not divisible by 10. Therefore, the postal service's scanners would detect an error, and the zip code would have to be read again.

1.1.2 *The Universal Product Code*

The Universal Product Code, or UPC, appears on virtually every item that we purchase. This is a 12 digit code consisting of two blocks of five digits preceded and followed with a single digit, as the barcode above indicates. The first six identify the country and the manufacturer of the product and the next five identify the product itself. The final digit is the check digit.

A 12 digit code $(a_1, \ldots, a_{12})$ is valid provided that

$$3a_1 + a_2 + 3a_3 + a_4 + \cdots + 3a_{11} + a_{12}$$

is evenly divisible by 10. The UPC of the example above is 0 41390 30860 4. Therefore, the sum for this code is

$$3\cdot 0+1\cdot 4+3\cdot 1+1\cdot 3+3\cdot 9+1\cdot 0+3\cdot 3+1\cdot 0+3\cdot 8+1\cdot 6\cdot +3\cdot 0+1\cdot 4=80.$$

This sum is indeed evenly divisible by 10, so the code is recognized as a valid UPC.

As with the zip code, given the first 11 digits, there is enough information to uniquely determine the check digit. For example, given the partial UPC of 0 71142 00001, if the check digit is x, then

$$3 \cdot 0 + 7 + 3 \cdot 1 + 1 + 3 \cdot 4 + 2 + 3 \cdot 0 + 0 + 3 \cdot 0 + 0 + 3 \cdot 1 + x = 28 + x,$$

which forces $x = 2$.

Just like the zip code scheme, the UPC scheme has the check digit to help detect errors. For example, if the code

0 7114200001 2

was incorrectly read as

0 7134200001 2

by reading the fourth digit as a 3 instead of as a 1, then the computation to check if this number is valid would give

$$3 \cdot 0 + 7 + 3 \cdot 1 + 3 + 3 \cdot 4 + 2 + 3 \cdot 0 + 0 + 3 \cdot 0 + +0 + 3 \cdot 1 + 2 = 32,$$

which is not divisible by 10. Therefore, a grocery store scanner would not recognize the code as valid, and the cashier would have to re-scan the item.

1.1.3 International Standard Book Numbers

Prior to 2009, books had been identified by a ten digit number, abbreviated by ISBN-10. For example, the book *Field and Galois Theory*, published by Springer, has for its ISBN-10 the number 0-387-94753-1. The first digit identifies the language in which the book is written, the second block of digits identifies the publisher, the third block identifies the book itself, and the final digit is the check digit. In this scheme, each digit can be a numeral $0, \ldots, 9$ or X, which represents 10. A ten digit number $(a_1, \ldots, a_{10})$ is a valid ISBN-10 provided that

$$10a_1 + 9a_2 + 8a_3 + \cdots + 2a_9 + a_{10}$$

is evenly divisible by 11. For the number above, we have

$$\begin{aligned} &10 \cdot 0 + 9 \cdot 3 + 8 \cdot 8 + 7 \cdot 7 + 6 \cdot 9 + 5 \cdot 4 + 4 \cdot 7 + 3 \cdot 5 + 2 \cdot 3 + 1 \\ &= 11 \cdot 24, \end{aligned}$$

so the number is indeed valid. The digit X is only used, when appropriate, for the check digit.

As with the previous two examples, the check digit can be determined uniquely, given that it is between 0 and 10. For example, for the book *A Classical Introduction to Modern Number Theory*, published by Springer, whose number will start with 0-387-97329, the check digit x must result in

$$\begin{aligned} &10 \cdot 0 + 9 \cdot 3 + 8 \cdot 8 + 7 \cdot 7 + 6 \cdot 9 + 5 \cdot 7 + 4 \cdot 3 + 3 \cdot 2 + 2 \cdot 9 + x \\ &= 265 + x \text{ divisible by } 11. \end{aligned}$$

Since $11 \cdot 24 = 264$, if $265 + x$ is to be divisible by 11 and $0 \leq x \leq 10$, then $x = 10$. Thus, the check digit for this book is X, and so the ISBN-10 is 0-387-97329-X.

The ISBN-10 scheme also allows for detection of some errors. When we discuss error detection in more detail, we will see that all of these schemes will detect an error in a single digit. On the other hand, errors in more than one digit are not always detected. However, the ISBN-10 scheme does better, in some sense, than the other two schemes above because it detects *transposition* errors. For example, given the ISBN 0-387-97329-X, if the fifth and sixth digits are transposed, the resulting number is 0-387-79329-X. The check for validity of this number would result in the sum

$$10 \cdot 0 + 9 \cdot 3 + 8 \cdot 8 + 7 \cdot 7 + 6 \cdot 7 + 5 \cdot 9 + 4 \cdot 3 + 3 \cdot 2 + 2 \cdot 9 + 10 = 273,$$

which is not divisible by 11. Thus, this number is invalid. In contrast, transposing digits in a valid zip code will always result in a number considered valid, since the sum of the digits is unchanged by this transposition and in UPC, while most transposition errors are detected, some are not.

In 2009, ISBN-10 was replaced by the 13 digit ISBN-13 check digit scheme which uses multipliers $(1, 3, 1, 3, 1, 3, 1, 3, 1, 3, 1, 3, 1)$ and divisibility by 10. Since ISBN-13 suffers the same deficiency with regard to transposition errors as does UPC, it seems that some error detection was sacrificed in favor of computational efficiency. We will discuss transposition errors in more detail later.

Exercises

1. Check whether or not the following numbers are valid ISBN-10s:
   (a) 0-8218-2169-5
   (b) 0-201-01361-9
   (c) 2-87647-089-6
   (d) 3-7643-3065-1
2. Suppose a UPC is read, but the third digit is left out, and the result is 0 $7x172$ 38175 1, where x represents the missing digit. Calculate, in terms of x, the sum needed to check if this is a valid number. Then write down the condition on x required for the number to be valid, and determine x.
3. The number 0-8176-3165-1 is an invalid ISBN-10 (Check this!). It was created by taking the ISBN-10 of a book and changing one digit. Can you tell which digit was changed? Explain why not by giving two examples of a valid ISBN-10 that differs from this one in exactly one digit.
4. Consider the following identification number scheme: If $a = (a_1, a_2, a_3, a_4)$, where each a_i is between 0 and 4, then the number a is valid provided that $4a_1 + 3a_2 + a_3 + 2a_4$ is divisible by 5. If $(3, 2, 4, x)$ is a valid number, determine x.
5. Consider the following identification number scheme: a valid number is a 5-tuple of integers $a = (a_1, a_2, a_3, a_4, a_5)$ with $0 \leq a_i \leq 12$ such that $2a_1 + 3a_2 + 5a_3 + a_4 + 6a_5$ is divisible by 13. If $(2, 3, 4, 11, x)$ is a valid number, determine x.
6. Consider the scheme of the previous problem. If $(a_1, a_2, a_3, a_4, a_5)$ is a valid number and if $a_1 \neq a_2$, prove that $(a_2, a_1, a_3, a_4, a_5)$ is not valid.

1.2 Modular Arithmetic

In order to investigate the error detection capabilities of the various identification number schemes we have discussed, and to work with the other applications in this course, we will look carefully at the computations involved in these schemes. In all three, a number is valid if some combination of its entries is divisible by some specific positive integer (10 or 11 in the examples). The actual result of the computation is not important in its own right. Rather what is important is only whether the result is divisible by the given integer. Phrased another way, what is important is not the numerical value of the computation, but rather the remainder we would get if we divide this number by our specific integer. In some sense we are doing arithmetic with these remainders when we do calculations in these schemes. While this may sound strange, we actually do it all the time.

Consider the following well-known scenario. When we tell time in the USA, the hour value is any whole number between 1 and 12. Three hours after 10 o'clock will be 1 o'clock. In general, to see what time it will be n hours after 10 o'clock, you add n to 10, and then remove enough multiples of 12 until you have a value between 1 and 12. For instance, in 37 hours past 10 o'clock, the time will be 11 o'clock since $47 = 36 + 11$. In telling time, we then identify 13 o'clock with 1 o'clock, 14 o'clock with 2 o'clock, and so on. In this clock arithmetic, if we add 12 hours to any time, we get the same time (but changing AM to PM and vice-versa). Therefore, 12 acts in clock arithmetic like 0 acts in ordinary arithmetic.

There is nothing special about 12 with respect to obtaining a new type of arithmetic. As we will see in more detail below, in doing calculations in the various identification number schemes we talked about above, we are essentially doing clock arithmetic, but with 12 replaced by 10 for the zip code and UPC, and by 11 for the ISBN-10 scheme. When we discuss coding theory, we will use clock arithmetic with 12 replaced by 2, and when we discuss cryptography, we will replace 12 by very large integers. We therefore need to discuss the general notion of clock arithmetic, more formally referred to as modular arithmetic.

We begin with a very familiar concept.

Definition 1.1. Let a and n be integers. We say that n divides a (or a is divisible by n) if $a = nb$ for some integer b.

Definition 1.2. Let n be a positive integer. We say that two integers a and b are congruent modulo n if $b - a$ is divisible by n. When this occurs, we write $a \equiv b \bmod n$.

Since $b-a$ is divisible by n exactly when $b-a = qn$ for some integer q, we see that $a \equiv b \bmod n$ if $b = a + qn$ for some q. This is a convenient way to express congruence modulo n in terms of an equation. If $n = 12$, then to say $a \equiv b \bmod 12$ is equivalent to saying a o'clock is the same time as b o'clock, if we ignore AM and PM. Congruence modulo n is a relation on the set of integers. The first thing we point out is that this relation is an equivalence relation.

Proposition 1.3. *The relation congruence modulo n is an equivalence relation for any positive integer n.*

Proof. Let n be a positive integer. We must prove that congruence modulo n is reflexive, symmetric, and transitive. For reflexivity, let a be any integer. Then $a \equiv a \bmod n$ since $a-a = 0$ is divisible by n; for $0 = n \cdot 0$. Next, for symmetry, suppose that a and b are integers with $a \equiv b \bmod n$. Then $b-a$ is divisible by n; say $b-a = qn$ for some integer q. Then $a-b = (-q)n$, so $a - b$ is also divisible by n. Therefore, $b \equiv a \bmod n$, and so this relation is symmetric.

Finally, to prove transitivity, suppose that a, b, c are integers with $a \equiv b \bmod n$ and $b \equiv c \bmod n$. By definition then, $b - a$ and $c - b$ are both divisible by n so that we may write

$$b - a = sn$$

$$c - b = tn$$

for some integers s, t. Adding these equations gives $c - a = (s + t)n$, so $c - a$ is divisible by n, and therefore $a \equiv c \bmod n$. This proves transitivity.

Since we have shown that congruence modulo n is reflexive, symmetric, and transitive, it is an equivalence relation. □

Understanding the equivalence classes of this relation is of crucial importance. Recall that if $\sim$ is an equivalence relation on a set X, then the equivalence class of an element $a \in X$ is the set $\{b \in X : b \sim a\}$. For ease of notation in the context of congruence modulo n, we shall write the equivalence class of an integer a by $\overline{a}$. Therefore,

$$\overline{a} = \{b \in \mathbb{Z} : b \equiv a \bmod n\}.$$

Suppose $n = 12$. The equivalence class of 0 consists of all integers that are congruent to 0 modulo 12. Since $12 \equiv 0 \bmod 12$ this equivalence class consists of all integers c with c o'clock equal to 12 o'clock. We have

$$\overline{0} = \{\dots, -24, -12, 0, 12, 24, \dots\}.$$

The equivalence class of 1 consists of all integers that are congruent to 1 modulo 12. That is, the equivalence class contains all integers c with c o'clock equal to 1 o'clock. We have

$$\overline{1} = \{\dots, -23, -11, 1, 13, 25, \dots\}.$$

Similarly, the equivalence class of 2 consists of all integers that are congruent to 2 modulo 12:

$$\overline{2} = \{\dots, -22, -10, 2, 14, 26, \dots\},$$

and so on for every positive integer.

Note that

$$\overline{12} = \{\dots, -12, 0, 12, \dots\}$$

contains 0, demonstrating that an equivalence class can be represented in different ways. We have $\overline{12} = \overline{0} = \overline{-24}$, and, more generally, $\overline{12} = \overline{12n}$ for any integer n. In other words, $\overline{12}$ is the equivalence class of any element of the set $\overline{12} = \{\dots, -12, 0, 12, \dots\}$, and similarly for $\overline{1}, \overline{2}$, etc.

If n is any positive integer, we denote by $\mathbb{Z}_n$ the set of equivalence classes of integers for the equivalence relation of congruence modulo n. For $n = 2$ we have

$$\mathbb{Z}_2 = \{\overline{0}, \overline{1}\}$$

where $\overline{1}$ represents the set of odd integers and $\overline{0}$ represents the set of even integers. Anybody who can tell time will see that

$$\mathbb{Z}_{12} = \left\{\overline{1}, \overline{2}, \overline{3}, \overline{4}, \overline{5}, \overline{6}, \overline{7}, \overline{8}, \overline{9}, \overline{10}, \overline{11}, \overline{12}\right\}.$$

Notice that while $\mathbb{Z}$ is an infinite set, $\mathbb{Z}_{12}$ is finite; even though each equivalence class is infinite, there are only finitely many of them! This is not special to 12; we will prove that $\mathbb{Z}_n$ has n elements for any n. We first need an important result that will prove useful in many places in this course. This result, known as the Division Algorithm, can be viewed as a formal statement of the process for writing a fraction as a proper fraction, i.e., the long division you learned in elementary school.

Theorem 1.4 (Division Algorithm). *Let a and n be integers with n positive. Then there are unique integers q and r with $a = qn + r$ and $0 \le r < n$.*

Proof. We use the *well-ordering* property of the integers that says any nonempty subset of the nonnegative integers has a smallest element. To use this property, let us define

$$S = \{s \in \mathbb{Z} : s \ge 0 \text{ and } s = a - qn \text{ for some } q \in \mathbb{Z}\}.$$

Note that $\mathcal{S}$ consists of all nonnegative remainders arising from divisions of a by n. First, we show that $\mathcal{S}$ is indeed nonempty. If $a \geq 0$, then $a - 0 \cdot n = a \in \mathcal{S}$. If $a < 0$, then $a - an = a(1-n) \geq 0$ since n is positive. Therefore, $a - an \in \mathcal{S}$. In either case, we see that $\mathcal{S}$ is nonempty. Therefore, by the well-ordering property, $\mathcal{S}$ contains a smallest element, which we call r. By definition of $\mathcal{S}$, there is a $q \in \mathbb{Z}$ with $r = a - qn$. Then $a = qn + r$, proving one part of the theorem. To show that $0 \leq r < n$, we note that $r \geq 0$ since $r \in \mathcal{S}$ and by definition $\mathcal{S}$ consists of nonnegative integers. If $r \geq n$, then $r - n = a - (q+1)n \in \mathcal{S}$ since $r - n \geq 0$. This would be a contradiction since $r - n$ is smaller than r. Therefore, $r < n$ as desired. Thus, we have produced integers q and r with $a = qn + r$ and $0 \leq r < n$. Next, we prove uniqueness of q and r. Suppose that q' and r' are a second pair of integers with $a = q'n + r'$ and $0 \leq r' < n$. Then $q'n + r' = qn + r$, so $(q' - q)n = r - r'$. Then $|q' - q|\, n = |r - r'|$. Since r and r' are both between 0 and $n - 1$, the absolute value of their difference is less than n. Since $|q' - q|\, n$ is a multiple of n, the only way the equation above can hold is if both sides are 0. Therefore, $r' = r$ and $q' = q$. This shows uniqueness of the integers q and r. □

In keeping with common terminology, the r above will be referred to as the *remainder* after dividing n into a. We can use the Division Algorithm to prove a simple but useful characterization of the relation congruence modulo n.

Lemma 1.5. *Let n be a positive integer and let a, b be integers. Then $a \equiv b \bmod n$ if and only if a and b have the same remainder after division by n. In other words, $a \equiv b \bmod n$ if and only if $a = qn + r$ and $b = q'n + r$ for some integers q and q'.*

Proof. Let n be a positive integer and a, b integers. Suppose that $a \equiv b \bmod n$. Then $b - a$ is divisible by n; say $b - a = tn$ for some integer t. By the Division Algorithm, we may write $a = qn + r$ and $b = q'n + s$ with q, q' integers, and r, s integers with $0 \leq r, s < n$. For concreteness, suppose that $r \leq s$. Then

$$\begin{aligned} tn &= b - a = q'n + s - (qn + r) \\ &= (q' - q)n + (s - r). \end{aligned}$$

Observe that $0 \leq s - r < n$, so the uniqueness assertion of the Division Algorithm applied to division of $b - a$ by n, shows that $t = q' - q$ and $0 = s - r$. Therefore $s = r$. The same conclusion is reached of course if we suppose that $s \leq r$. This shows that a and b have the same remainder after division by n. Conversely, suppose that $a = qn + r$ and $b = q'n + r$ for some integers q, q', r. Then

$$b - a = q'n + r - (qn + r) = (q' - q)n,$$

so $b - a$ is divisible by n. Therefore, $a \equiv b \bmod n$. □

Corollary 1.6. *Let n be a positive integer. Every integer is congruent modulo n to exactly one integer between 0 and $n - 1$.*

Proof. Let n be a positive integer. If a is an integer, then the Division Algorithm gives us integers q and r with $a = qn + r$ and $0 \leq r < n$. Thus, $a - r = qn$ is divisible by n, so $a \equiv r \bmod n$. If s is between 0 and $n - 1$ and $a \equiv s \bmod n$, then the previous lemma shows us that a and s have the same remainders after division by n. But, since $s = 0 \cdot n + s$ and $a = qn + r$, the lemma tells us that $s = r$. Thus, a is congruent modulo n to exactly one integer between 0 and $n - 1$. □

Definition 1.7. Let n be a positive integer and a an arbitrary integer. The unique integer between 0 and $n-1$ to which a is congruent modulo n is called the least residue of a modulo n.

Corollary 1.8. *For every positive integer n, the cardinality $|\mathbb{Z}_n|$ of $\mathbb{Z}_n$ is equal to n, and* $\mathbb{Z}_n = \{\overline{0}, \overline{1}, \ldots, \overline{n-1}\}$.

Proof. According to Corollary 1.6, the equivalence classes of integers modulo n are in 1-1 correspondence with the remainders after division by n. These remainders are the precisely the integers in the set $\{0, 1, \ldots, n-1\}$, hence the corollary. □

Exercises

1. Let n be a positive integer and let m be a positive divisor of n. If a and b are integers with $a \equiv b \bmod n$, prove that $a \equiv b \bmod m$.
2. For each nonnegative integer i, what is the least residue modulo 9 of 10^i?
3. Starting with $i = 0$ find the first few values of the least residue modulo 3 of 4^i. Formulate a conjecture about these values.
4. Prove for each integer n, that n^2 is congruent either to 0 or to 1 modulo 4.
5. Calculate the least residues modulo 5 of $1^4, 2^4, 3^4$, and 4^4.
6. Calculate the least residues modulo 7 of $1^6, 2^6, 3^6, 4^6, 5^6$, and 6^6.

1.2.1 Arithmetic Operations in $\mathbb{Z}_n$

We now discuss a generalization of clock arithmetic for any modulus n. Recall that to determine what time it will be 7 hours after 9 o'clock, we add $7+9$, getting 16, then subtract 12 to get 4, and conclude that the time will be 4 o'clock. In other words, we add the numbers and then subtract enough multiples of 12 to get a valid time. This is the idea behind addition in $\mathbb{Z}_n$. Similarly, we can define multiplication.

Definition 1.9. Let n be a positive integer and let $\overline{a}$ and $\overline{b}$ be elements of $\mathbb{Z}_n$. Then $\overline{a} + \overline{b} = \overline{a+b}$ and $\overline{a} \cdot \overline{b} = \overline{ab}$.

What the definition tells is that to add two elements of $\mathbb{Z}_n$, we represent them as the equivalence class of some integers, then we add the integers, then take the equivalence class of the sum. Similarly, to multiply two elements of $\mathbb{Z}_n$ multiply two integer representatives and take the equivalence class modulo n of the product. Working in $\mathbb{Z}_{12}$ for example, the definition of addition yields

$$\begin{aligned} \overline{7} + \overline{9} &= \overline{16} \\ &= \overline{4} \end{aligned}$$

and

$$\begin{aligned} \overline{7} \cdot \overline{9} &= \overline{63} \\ &= \overline{3}. \end{aligned}$$

Applying these definitions, we obtain addition and multiplication tables for $\mathbb{Z}_2$:

$+ \bmod 2$	$\overline{0}$	$\overline{1}$
$\overline{0}$	$\overline{0}$	$\overline{1}$
$\overline{1}$	$\overline{1}$	$\overline{0}$

$\cdot \bmod 2$	$\overline{0}$	$\overline{1}$
$\overline{1}$	$\overline{0}$	$\overline{0}$
$\overline{1}$	$\overline{0}$	$\overline{1}$

The addition and multiplication tables for $\mathbb{Z}_6$ are as follows:

$+ \bmod 6$	$\overline{0}$	$\overline{1}$	$\overline{2}$	$\overline{3}$	$\overline{4}$	$\overline{5}$
$\overline{0}$	$\overline{0}$	$\overline{1}$	$\overline{2}$	$\overline{3}$	$\overline{4}$	$\overline{5}$
$\overline{1}$	$\overline{1}$	$\overline{2}$	$\overline{3}$	$\overline{4}$	$\overline{5}$	$\overline{0}$
$\overline{2}$	$\overline{2}$	$\overline{3}$	$\overline{4}$	$\overline{5}$	$\overline{0}$	$\overline{1}$
$\overline{3}$	$\overline{3}$	$\overline{4}$	$\overline{5}$	$\overline{0}$	$\overline{1}$	$\overline{2}$
$\overline{4}$	$\overline{4}$	$\overline{5}$	$\overline{0}$	$\overline{1}$	$\overline{2}$	$\overline{3}$
$\overline{5}$	$\overline{5}$	$\overline{0}$	$\overline{1}$	$\overline{2}$	$\overline{3}$	$\overline{4}$

$\cdot \bmod 6$	$\overline{0}$	$\overline{1}$	$\overline{2}$	$\overline{3}$	$\overline{4}$	$\overline{5}$
$\overline{0}$	$\overline{0}$	$\overline{0}$	$\overline{0}$	$\overline{0}$	$\overline{0}$	$\overline{0}$
$\overline{1}$	$\overline{0}$	$\overline{1}$	$\overline{2}$	$\overline{3}$	$\overline{4}$	$\overline{5}$
$\overline{2}$	$\overline{0}$	$\overline{2}$	$\overline{4}$	$\overline{0}$	$\overline{2}$	$\overline{4}$
$\overline{3}$	$\overline{0}$	$\overline{3}$	$\overline{0}$	$\overline{3}$	$\overline{0}$	$\overline{3}$
$\overline{4}$	$\overline{0}$	$\overline{4}$	$\overline{2}$	$\overline{0}$	$\overline{4}$	$\overline{2}$
$\overline{5}$	$\overline{0}$	$\overline{5}$	$\overline{4}$	$\overline{3}$	$\overline{2}$	$\overline{1}$

For alternative views of addition and multiplication, recall that the equivalence classes of 7 and of 9 mod 12 are:

$$\overline{7} = \{\ldots, -5, 7, 19, \ldots\},$$
$$\overline{9} = \{\ldots, -3, 9, 21, \ldots\}.$$

A new set of integers is obtained by taking all possible sums of the integers in $\overline{7}$ with the integers in $\overline{9}$. This set is $\{\ldots, -8, 4, 16, 28, \ldots\}$, which is precisely $\overline{4}$. We can view addition of equivalence classes as this method of adding sets of integers together. Similarly, the set of all products of the integers in $\overline{7}$ with the integers in $\overline{9}$, namely $\{\ldots, 15, -45, -105, -21, 63, \ldots\}$, all of which reduce to 3 modulo 12.

The definitions above allow us to do modular arithmetic more simply than these set operations. However, there is one problem with the definition. When we write an equivalence class as $\overline{a}$, this is describing the class by one particular member a of it. The choice of a is not unique. For example, still in $\mathbb{Z}_{12}$, we have $\overline{7} = \overline{-5} = \overline{91}$, giving three ways to describe the same class. Similarly, $\overline{9} = \overline{21}$. The problem is this: If we use different representations of two equivalence classes, do we get the same result when we add or multiply? If the answer is no, then we have a meaningless definition. Therefore, we need to verify that our definition is valid. For example, we have $\overline{7} + \overline{9} = \overline{16} = \overline{4}$, and $\overline{91} + \overline{21} = \overline{112} = \overline{4}$, since $112 = 9 \cdot 12 + 4$. That the consistency we see in this example holds in general for modular addition and multiplication is explained by the following lemma.

Lemma 1.10. *Let n be a positive integer. If a, b, c, d are integers with $a \equiv c \bmod n$ and $b \equiv d \bmod n$, then $a + b \equiv c + d \bmod n$ and $ab \equiv cd \bmod n$.*

Proof. Let n be a positive integer, and suppose that $a \equiv c \bmod n$ and $b \equiv d \bmod n$. Then $c - a$ and $d - b$ are divisible by n, so there are integers s, t with $a - c = sn$ and $b - d = tn$. Thus $a = c + sn$ and $b = d + tn$. By adding the equations, we get

$$a + b = (c + sn) + (d + tn) = (c + d) + (s + t)n,$$

which shows that $(a + b) - (c + d)$ is a multiple of n. Therefore, $a + b \equiv c + d \bmod n$. If we multiply both equations, we get

$$\begin{aligned} ab &= (c + sn)(d + tn) = cd + ctn + snd + sntn \\ &= cd + (ct + sd + snt)n, \end{aligned}$$

so $ab - cd$ is a multiple of n. Thus, $ab \equiv cd \bmod n$. This proves the lemma. □

Lemma 1.10 tells us that Definition 1.9 is meaningful, i.e., that arithmetic modulo n is well defined. Because of the simple formula for addition and multiplication in $\mathbb{Z}_n$, we obtain analogues of many of the common properties of integer arithmetic. In particular, the following properties hold for all elements $\overline{a}, \overline{b}, \overline{c}$ of $\mathbb{Z}_n$:

- Commutativity of Addition: $\overline{a} + \overline{b} = \overline{b} + \overline{a}$;
- Associativity of Addition: $(\overline{a} + \overline{b}) + \overline{c} = \overline{a} + (\overline{b} + \overline{c})$;
- Existence of an Additive Identity: $\overline{a} + \overline{0} = \overline{a}$;
- Existence of Additive Inverses: $\overline{a} + \overline{-a} = \overline{0}$;
- Commutativity of Multiplication: $\overline{a} \cdot \overline{b} = \overline{b} \cdot \overline{a}$;
- Associativity of Multiplication: $(\overline{a} \cdot \overline{b}) \cdot \overline{c} = \overline{a} \cdot (\overline{b} \cdot \overline{c})$;
- Existence of a Multiplicative Identity: $\overline{a} \cdot \overline{1} = \overline{a}$;
- Distributivity: $\overline{a} \cdot (\overline{b} + \overline{c}) = \overline{a} \cdot \overline{b} + \overline{a} \cdot \overline{c}$.

While we will not write out proofs for all these properties, we give the idea of how to prove them with one example. For commutativity of addition, we have

$$\overline{a} + \overline{b} = \overline{a + b} = \overline{b + a} = \overline{b} + \overline{a}.$$

Note that we used the definition of addition in $\mathbb{Z}_n$ twice. The only other property used was the familiar commutativity of addition in $\mathbb{Z}$. Every other property in the list above comes from a combination of the definition of addition and/or multiplication and the corresponding properties of these operations in integer arithmetic.

Subtraction in $\mathbb{Z}_n$ is defined by $\overline{a} - \overline{b} = \overline{a - b}$. Another way to write subtraction is by $\overline{a} - \overline{b} = \overline{a} + \overline{-b}$. Because of the fourth property in the list above, $\overline{-b}$ is the additive inverse of $\overline{b}$, and the subtraction $\overline{a} - \overline{b}$ is the same as the sum of $\overline{a}$ and the additive inverse $\overline{-b}$ of $\overline{b}$, just as the case of the real numbers.

There are some differences between arithmetic in $\mathbb{Z}_n$ and ordinary arithmetic. In $\mathbb{Z}$, if two integers a and b satisfy $ab = 0$, then either $a = 0$ or $b = 0$. However, this is not always true in $\mathbb{Z}_n$. For example, in $\mathbb{Z}_{10}$, we have $\overline{2} \cdot \overline{5} = \overline{10} = \overline{0}$. In $\mathbb{Z}_{12}$ we have $\overline{8} \cdot \overline{3} = \overline{24} = \overline{0}$. In contrast though, in $\mathbb{Z}_{11}$, one can show (and we leave it for a homework exercise) that if $\overline{a} \cdot \overline{b} = \overline{0}$, then $\overline{a} = \overline{0}$ or $\overline{b} = \overline{0}$. We will see shortly that this has consequences for detection of errors in the identification number schemes we discussed earlier. In particular, the ISBN-10 scheme, in which 11 has a special role, can detect certain types of errors that remain undetected by the zip code and UPC schemes, both of which utilize 10.

Another difference has to do with division. If we restrict ourselves to $\mathbb{Z}$, the only time we can solve the equation $ab = 1$ is with $a = b = 1$ or $a = b = -1$. In $\mathbb{Z}_n$, we have the corresponding solutions to the equation $\overline{a} \cdot \overline{b} = \overline{1}$. However, we may have more solutions. For example, in $\mathbb{Z}_{10}$, we have $\overline{3} \cdot \overline{7} = \overline{21} = \overline{1}$ and in $\mathbb{Z}_{11}$ we have $\overline{8} \cdot \overline{7} = \overline{56} = \overline{1}$. In fact, for $\overline{a} \in \mathbb{Z}_{11}$ with $\overline{a} \neq \overline{0}$, there is a $\overline{b} \in \mathbb{Z}_{11}$ with $\overline{a} \cdot \overline{b} = \overline{1}$. This is also left to a homework exercise. As we will see below, the fact that $\overline{3} \cdot \overline{x} = \overline{1}$ can be solved in $\mathbb{Z}_{10}$ was crucial in the decision to use the vector $(3, 1, \ldots, 3, 1)$ in the UPC scheme.

1.2.2 Greatest Common Divisors

To facilitate our discussion of error detection, and for future applications, we need some facts about the greatest common divisor of two integers. Later we will see that everything we do here will have analogues for polynomials. Those results will be crucial for our treatments of ruler and compass constructions and of error correcting codes.

Definition 1.11. Let a and b be integers, at least one of which is nonzero. Then the greatest common divisor of a and b, denoted by $\gcd(a, b)$, is the largest integer that divides both a and b.

The greatest common divisor of any pair (a, b) of integers exists provided at least one of them is nonzero. The reason for the condition that at least one of a or b is the simple fact that every integer divides 0. Thus if $a = b = 0$, then every integer divides both a and b and no largest common divisor exists. If, on the other hand, $a \neq 0$, then each divisor of a is no larger than $|a|$ so the set of common divisors of a and b is bounded above by the larger of $|a|$ and $|b|$. In particular a largest common divisor must exist. The notation $\gcd(a, b)$ is used only under the condition that at least one of a or b is nonzero. Clearly if $a \neq 0$, then $\gcd(a, 0) = |a|$. More generally, if the integer d divides two integers, then so does $|d|$ so that $\gcd(a, b) \geq 1$. For example, $\gcd(4, 6) = 2$ and $\gcd(-20, 24) = 4$. Of course $\gcd(a, b)$ can equal 1, e.g. $\gcd(3, 10) = 1$ and $\gcd(1, b) = 1$ for every b.

Proposition 1.13 describes one of the most useful properties of the greatest common divisor of integers a and b. This result is a consequence of the Division Algorithm, and gives an important representation of the greatest common divisor. To help with the proof, we prove the following lemma.

Lemma 1.12. *Let a and b be integers, and suppose c is an integer that divides both a and b. Then c divides $ax + by$ for any integers x and y.*

Proof. Suppose c divides a and b, and let x and y be arbitrary integers. Since c divides a and c divides b there are integers α and β with $a = \alpha c$ and $b = \beta c$. Then

$$ax + by = \alpha cx + \beta cy = c(\alpha x + \beta y).$$

Since $\alpha x + \beta y$ is an integer, this equation shows that c divides $ax + by$. □

For given integers a, b, expressions of the form $ax + by$ with x and y integers are called integer linear combinations of a and b. More generally, given integers $a_1, a_2, \dots a_n$, an expression of the form $\sum_{i=1}^{n} a_i x_i$, with integers $x_1, x_2, \dots x_n$ is called an integer linear combination of $a_1, a_2, \dots a_n$.

Proposition 1.13. *Let a and b be integers, at least one of which is nonzero, and set $d = \gcd(a, b)$. Then d can be expressed as an integer linear combination of a and b, (i.e., there are integers x and y with $d = ax + by$).*

Proof. To prove this we use an argument reminiscent of that used to prove the Division Algorithm. Let

$$\mathcal{S} = \{as + bt : s, t \in \mathbb{Z}, as + bt > 0\}.$$

Once $\mathcal{S}$ is shown to be nonempty, we take its least element, whose existence is guaranteed by the well-ordering property, and prove that it is the greatest common divisor of a and b. To see that $\mathcal{S}$ is nonempty, note that $a^2 + b^2$ is positive and also a linear combination of a and b, hence an element of $\mathcal{S}$.

Now let e be the least element of $\mathcal{S}$. By the defining property of $\mathcal{S}$ there are integers x and y with $e = ax + by$. Set $d = \gcd(a, b)$ and we proceed to show that $e = d$. First, since d divides

a and b, Lemma 1.12 implies that d divides $e = ax + by$. Since e and d are both positive, this forces $d \le e$. We show the reverse inequality by showing that e is a common divisor of a and b. From the Division Algorithm, we may write $a = qe + r$ with q and r integers and $0 \le r < e$. Then

$$\begin{aligned} r &= a - qe = a - q(ax + by) \\ &= a(1 - qx) + b(-qy). \end{aligned}$$

If $r > 0$, this equation would show that $r \in \mathcal{S}$. This is impossible since $r < e$ and e is the least element of $\mathcal{S}$. Therefore $r = 0$, which means that $a = qe$, so e divides a. The identical argument, with a replaced by b shows that e divides b as well, forcing $e \le d$ because d is the greatest of the common divisors of a and b. Since $e \le d$ and $d \le e$, we get $e = d$. Therefore we have written $d = e = ax + by$, as desired. □

Corollary 1.14. *Let a and b be integers, not both zero, and let $d = \gcd(a, b)$. If c is any common divisor of a and b, then c divides d.*

Proof. By the proposition, we may write $d = ax + by$ for some integers x, y. If c is a common divisor of a and b, then c divides $d = ax + by$ by the lemma. □

Recall that a positive integer $p > 1$ is *prime* if the only positive divisors of p are 1 and p. It follows immediately that if a is any integer, then $\gcd(a, p) = 1$ or $\gcd(a, p) = p$. Since the latter equality holds only if p divides a, we see that $\gcd(a, p) = 1$ for any integer a not divisible by p. It is quite common for the condition $\gcd(a, b) = 1$ with neither a nor b prime. For example, $\gcd(9, 16) = 1$ and $\gcd(10, 21) = 1$. This condition is important enough for us to study it.

Definition 1.15. Two integers a and b are said to be relatively prime if $\gcd(a, b) = 1$.

In light of this definition, Proposition 1.13 also has the following consequence.

Corollary 1.16. *If a and b are integers, not both zero, then $\gcd(a, b) = 1$ if and only if there are integers x and y with $1 = ax + by$.*

Proof. One direction follows immediately from the proposition: if $\gcd(a, b) = 1$, then $1 = ax + by$ for some integers x and y. For the converse, suppose there are integers x and y with $1 = ax + by$. We wish to show that $\gcd(a, b) = 1$. Let $d = \gcd(a, b)$. By Lemma 1.12, d divides $1 = ax + by$. However, d is a positive integer, and the only positive integer that divides 1 is 1 itself. Therefore, $d = 1$. □

A nice consequence of this corollary, in the terminology of modular arithmetic, is the following result.

Corollary 1.17. *Let n be a positive integer. The equation $\overline{a} \cdot x = \overline{1}$ has a solution in $\mathbb{Z}_n$ if and only if $\gcd(a, n) = 1$.*

Proof. Let n be a positive integer and a be an integer with $\gcd(a, n) = 1$. Then there are integers s and t with $as + nt = 1$. Rewriting this, we have $as = 1 + n(-t)$, which says that $as \equiv 1 \bmod n$, or $\overline{as} = \overline{1}$ in $\mathbb{Z}_n$. Thus, we have produced a solution to the equation $\overline{a}x = \overline{1}$ in $\mathbb{Z}_n$.

Conversely, if $\overline{as} = \overline{1}$ in $\mathbb{Z}_n$, then $as = 1 + n(-t)$ for some integer t. Thus $as + nt = 1$ and Corollary 1.16 yields that $\gcd(a, n) = 1$. □

Caution: Corollary 1.16 generalizes to the case in which $\gcd(a, b) > 1$, but not perhaps in the most obvious way. That some seemingly small integer $c > 1$ can be expressed as a linear combination of a and b does not guarantee $c = \gcd(a, b)$. To be more specific, let $a = 10$ and $b = 7$. Then

$2 = 10 \cdot (3) + 7 \cdot (-4)$, but $\gcd(10, 7) = 1 = 10(-2) + 7(3)$. Indeed, if a and b are not both zero, Proposition 1.13 realizes $\gcd(a, b)$ as the smallest positive integer linear combination of them.

Although the following very important result won't be used in the discussion of error detection capability of identification number schemes, it will arise in other applications.

Proposition 1.18. *Let a, b, and c be integers such that a divides bc. If $\gcd(a, b) = 1$, then a divides c.*

Proof. Suppose a divides bc and $\gcd(a, b) = 1$. The first condition implies that there is an integer α with $bc = \alpha a$. The second condition and the previous proposition shows that there are integers x, y with $1 = ax + by$. Multiplying this equation by c yields $c = axc + bcy$, and so $c = axc + \alpha ay$ by substituting the equation $bc = \alpha a$. Therefore, $c = a(xc + \alpha y)$. Since $xc + \alpha y$ is an integer, a divides c, as desired. □

1.2.3 The Euclidean Algorithm

In most applications, the method of choice for computing the greatest common divisor of two integers is the Euclidean Algorithm. In principle, prime factorization can be used to compute the greatest common divisor, but if the numbers are large, factorization becomes very time consuming. That fact turns out to be the basis of the security of the RSA scheme for cryptography discussed in Chap. 8. The Euclidean Algorithm has the advantage of computational ease; Maple and other computer programs use this algorithm for greatest common divisor computations.

The Euclidean Algorithm to calculate $\gcd(a, b)$ for two positive integers a and b consists of the following steps:

1. Set $a_0 = \max(a, b)$ and $a_1 = \min(a, b)$. Set $i = 1$.
2. If $a_i = 0$, then $a_{i-1} = \gcd(a, b)$.
3. If $a_i \neq 0$, then divide a_i into a_{i-1}, getting $a_{i-1} = qa_i + r$.
4. Replace i by $i + 1$.
5. Define $a_i = r$.
6. Go to Step 2.

The Euclidean Algorithm employs the Division Algorithm iteratively to produce a strictly decreasing sequence of remainders. Specifically, with b the larger of the two integers, divide a into b and find the remainder. Then divide this remainder into a, finding the second remainder. Then divide the second remainder into the first remainder, to obtain the third remainder. According to the Division Algorithm each successive remainder is strictly less than its predecessor. But, since all the remainders are nonnegative, eventually one of them is equal to 0, so that the algorithm terminates after some finite stage. As we will see, the final nonzero remainder must be the greatest common divisor. Before proving this fact, we consider some examples.

Example 1.19. To perform this algorithm to calculate $\gcd(10, 14)$, we list the necessary iterations.

- Set $a_0 = 14$ and $a_1 = 10$. We also start with $i = 1$. We have $14 = 1 \cdot 10 + 4$. Set $i = 2$ and $a_2 = 4$. Return to Step 2 and note that $a_2 \neq 0$.
- We have $a_1 = 10$ and $a_2 = 4$ and we see that $10 = 2 \cdot 4 + 2$. Set $i = 3$ and $a_3 = 2$. Return to Step 2 and note that $a_3 \neq 0$.
- We have $a_2 = 4$ and $a_3 = 2$. Write $4 = 2 \cdot 2 + 0$. Set $i = 4$ and $a_4 = 0$.
- As $a_3 = 2$ and $a_4 = 0$, Step 2 then yields $a_3 = \gcd(10, 14)$.

Example 1.20. Here we streamline the process and calculate $\gcd(12342, 2738470)$, listing the divisions and remainders encountered.

$$\begin{aligned}
2738470 &= 221 \cdot 12342 + \mathbf{10888} \\
12342 &= 1 \cdot \mathbf{10888} + \mathbf{1454} \\
10888 &= 7 \cdot \mathbf{1454} + \mathbf{710} \\
1454 &= 2 \cdot \mathbf{710} + \mathbf{34} \\
710 &= 20 \cdot \mathbf{34} + \mathbf{30} \\
34 &= 1 \cdot \mathbf{30} + \mathbf{4} \\
30 &= 7 \cdot \mathbf{4} + \mathbf{2} \\
4 &= 2 \cdot \mathbf{2} + \mathbf{0}.
\end{aligned}$$

The last nonzero remainder is 2, so $2 = \gcd(12342, 2738470)$.

Example 1.21. To calculate $\gcd(849149, 9889)$ we do the following arithmetic:

$$\begin{aligned}
849149 &= 85 \cdot \mathbf{9889} + 8584 \\
9889 &= 1 \cdot \mathbf{8584} + 1305 \\
8584 &= 6 \cdot \mathbf{1305} + 754 \\
1305 &= 1 \cdot \mathbf{754} + 551 \\
754 &= 1 \cdot \mathbf{551} + 203 \\
551 &= 2 \cdot \mathbf{203} + 145 \\
203 &= 1 \cdot \mathbf{145} + 58 \\
145 &= 2 \cdot \mathbf{58} + 29 \\
58 &= 2 \cdot \mathbf{29} + 0
\end{aligned}$$

Since the final nonzero remainder is 29, we have $29 = \gcd(849149, 9889)$.

That the Euclidean Algorithm works is based on two facts. First, if a is any nonzero positive integer, then $\gcd(a, 0) = a$. This is clear from the definition since a is clearly the largest integer that divides a, and every nonzero integer divides 0. Second, and more important, is the following result.

Lemma 1.22. *Suppose that a and b are integers, not both zero, with $b = qa + r$ for some integers q and r. Then $\gcd(a, b) = \gcd(a, r)$.*

Proof. Let $d = \gcd(a, b)$ and $e = \gcd(a, r)$. From $b = qa + r$ we see that e divides b so that $e \le d$. But $r = b - qa$ so that d divides r and $d \le e$ as well, forcing $e = d$. □

What this lemma says is that to calculate $\gcd(a, b)$, we can divide a into b, and then replace b by the remainder r. The Euclidean Algorithm uses this repeatedly until a remainder of zero is obtained, then the previous fact is applied. To make this more explicit, we work one more example.

Example 1.23. To find $\gcd(24, 112)$, we list both the calculations and what the calculation yields.

$$\begin{aligned} 112 &= 4 \cdot 24 + 16 & \gcd(24, 112) &= \gcd(24, 16) \\ 24 &= 1 \cdot 16 + 8 & \gcd(24, 16) &= \gcd(16, 8) \\ 16 &= 2 \cdot 8 + 0 & \gcd(16, 8) &= \gcd(8, 0) = 8 \end{aligned}$$

The algorithm actually does more than just compute greatest common divisors. It also provides an algorithm to express $\gcd(a, b)$ as the linear combination guaranteed by Proposition 1.13. To illustrate this, we revisit the previous example. We saw that $\gcd(24, 112) = 8$. The second to last equation is $24 = 1 \cdot 16 + 8$; equivalently, solving for the remainder term,

$$8 = 1 \cdot 24 + (-1) \cdot 16.$$

Thus, we have written $8 = \gcd(24, 112)$ as a linear combination of 24 and 16. The equation before that (the first) is $112 = 4 \cdot 24 + 16$ and, solving for the remainder term here,

$$16 = 1 \cdot 112 + (-4) \cdot 24.$$

Replacing 16 in the previous displayed equation by this expression yields

$$\begin{aligned} 8 &= 1 \cdot 24 + (-1) \cdot (1 \cdot 112 + (-4) \cdot 24) \\ &= 1 \cdot 24 - 1 \cdot 112 + 4 \cdot 24 \\ &= 5 \cdot 24 + (-1) \cdot 112, \end{aligned}$$

which is the desired representation. Note how the coefficients of 24 and 112 were collected in the last step.

In general, to recover $\gcd(a, b)$ as a linear combination of a and b, record all of the steps in the Euclidean Algorithm that produce $\gcd(a, b)$. Then, as in the example, view the equation in which the greatest common divisor arises as the last nonzero remainder as an expression of the greatest common divisor as a linear combination of the two previous remainders. Then work up the equations, successively substituting for the remainder term arising in each step. Then rewrite this equation as an expression of the greatest common divisor as a linear combination either of previous remainders or the initial data a and b (when the very first equation is reached).

Example 1.24. To express $1 = \gcd(8, 29) = 8x + 29y$, first apply the Euclidean Algorithm:

$$\begin{aligned} 29 &= 3 \cdot 8 + \mathbf{5} & (1) \\ 8 &= 1 \cdot 5 + \mathbf{3} & (2) \\ 5 &= 1 \cdot 3 + \mathbf{2} & (3) \\ 3 &= 1 \cdot 2 + \mathbf{1} & (4). \end{aligned}$$

Next, beginning with the last equation, work up, substituting for remainders and collecting coefficients of the remainder that arose in the previous equation:

$$\begin{aligned} \mathbf{1} &= 3 - 1 \cdot \mathbf{2} & (4) \\ &= 3 - 1 \cdot (\mathbf{5} - \mathbf{3}) & (3) \\ &= 2 \cdot \mathbf{3} - 1 \cdot \mathbf{5} & \\ &= 2 \cdot (\mathbf{8} - \mathbf{5}) - 1 \cdot 5 & (2) \\ &= 2 \cdot 8 - 3 \cdot 5. & \end{aligned}$$

Finally,

$$
\begin{aligned}
1 &= 2 \cdot 8 - 3(\mathbf{29} - \mathbf{3} \cdot \mathbf{8}) \quad (1) \\
&= 11 \cdot 8 - 3 \cdot 29.
\end{aligned}
$$

The final result, $1 = 11 \cdot 8 - 3 \cdot 29$, expresses $1 = \gcd(8, 29)$ as the desired linear combination, with $x = 8, y = -3$.

Exercises

1. For which $\overline{a}$ in $\mathbb{Z}_{10}$ does the equation $\overline{a} \cdot \overline{x} = \overline{0}$ have only the trivial solution $\overline{x} = \overline{0}$? When there is a nontrivial solution, give all solutions.
2. For which $\overline{a}$ in $\mathbb{Z}_{12}$ does the equation $\overline{a} \cdot \overline{x} = \overline{0}$ have only the trivial solution $\overline{x} = \overline{0}$? When there is a nontrivial solution, give all solutions.
3. For which $\overline{a}$ in $\mathbb{Z}_{11}$ does the equation $\overline{a} \cdot \overline{x} = \overline{0}$ have only the trivial solution $\overline{x} = \overline{0}$? When there is a nontrivial solution, give all solutions.
4. Let n be a positive integer. Based on the previous problems, come up with a precise conjecture for which $\overline{a}$ in $\mathbb{Z}_n$ does the equation $\overline{a} \cdot \overline{x} = \overline{0}$ have only the trivial solution $\overline{x} = \overline{0}$. State your reasoning for coming up with your conjecture. Prove your conjecture.
5. For which $\overline{a}$ in $\mathbb{Z}_{10}$ does the equation $\overline{a} \cdot \overline{x} = \overline{1}$ have a solution? For each $\overline{a}$ for which the equation has a solution, state the solution.
6. In $\mathbb{Z}_{11}$, verify that the equation $\overline{a} \cdot \overline{x} = \overline{1}$ has a solution for every $\overline{a} \neq \overline{0}$.
7. Solve the equation $\overline{4} \cdot \overline{x} = \overline{2}$ in $\mathbb{Z}_6$. Can you solve the equation $\overline{4} \cdot \overline{x} = \overline{3}$ in $\mathbb{Z}_6$? Why or why not?
8. Solve the equation $\overline{e} \cdot \overline{x} = \overline{1}$ in $\mathbb{Z}_n$, where $n = 7325494815531218239807$ and $e = 1977326753$. If you do not wish to do this by hand, the values of n and e are input in the Maple worksheet Section-1.2-Exercise-8.mw. Read fully that worksheet and follow the instructions in it.

   (We will perform this type of calculation, except with much larger values of n, when we talk about cryptography.)
9. Prove the distributive law in $\mathbb{Z}_n$: if $\overline{a}$, $\overline{b}$, and $\overline{c}$ are arbitrary elements of $\mathbb{Z}_n$, then $\overline{a} \cdot (\overline{b} + \overline{c}) = \overline{a} \cdot \overline{b} + \overline{a} \cdot \overline{c}$. State explicitly which facts and/or definitions you use in your proof.
10. If p and q are distinct prime numbers, prove that p and q are relatively prime.
11. Explain, with the help of Lemma 1.22, why the Euclidean Algorithm produces the gcd of two integers.
12. Apply the Euclidean Algorithm, by hand, to find $\gcd(132, 50)$ and to express it as a linear combination of the two given integers. **Write out all the steps you take.**
13. Use your calculation in the previous problem to solve to the congruence $50x \equiv 4 \bmod 132$. Again, do this by hand.
14. Let $n = 7325494815531218239807$ and $e = 1977326753$. Calculate $\gcd(e, n)$, write this gcd as a linear combination of e and n, and use this data to solve the equation $ex \equiv 1 \bmod n$. Feel free to use Maple; the file Section-1.2-Exercise-14.mw has the values of e and n along with a reminder of which Maple command will help you to do this.
15. Suppose we have an unknown integer x such that $46+3x$ is a multiple of 10. Give the justification for each step of the following calculation in $\mathbb{Z}_{10}$. Note that some steps require more than one property of $\mathbb{Z}_{10}$.

$$\overline{46 + 3x} = \overline{0}$$
$$\overline{46} + \overline{3} \cdot \overline{x} = \overline{0}$$
$$\overline{3} \cdot \overline{x} = -\overline{46}$$
$$\overline{3} \cdot \overline{x} = \overline{4}$$
$$\overline{7} \cdot (\overline{3} \cdot \overline{x}) = \overline{7} \cdot \overline{4} = \overline{8}$$
$$\overline{x} = \overline{8}.$$

16. Let n be a positive integer. If a and b are integers with $a \equiv b \bmod n$, prove that $a^t \equiv b^t \bmod n$ for all positive integers t.
17. Suppose that a, b, n, s are integers with $as \equiv b \bmod n$. Prove that $\gcd(a, n)$ divides b.
18. Suppose that a, b, n are integers such that $\gcd(a, n)$ divides b. Prove that $ax \equiv b \bmod n$ has a solution. Use the linear combination representation of the gcd to find a solution to the equation $10x \equiv 4 \bmod 22$.

    (Hint: Write $d = \gcd(a, n)$ as a linear combination of a and n and write $b = qd$ for some integer q. Manipulate these two equations.)
19. Let n be a positive integer. If a is an integer with $\gcd(a, n) > 1$, prove that there is a nonzero $\overline{b}$ in $\mathbb{Z}_n$ with $\overline{a} \cdot \overline{b} = \overline{0}$ in $\mathbb{Z}_n$. Conclude that the equation $\overline{a} \cdot x = \overline{1}$ cannot be solved in $\mathbb{Z}_n$.
20. Prove the divisibility test for 9: A number, written in normal base 10 form, is divisible by 9 if and only if the sum of its digits is divisible by 9.
21. Prove the divisibility test for 11: A number, written in normal base 10 form, is divisible by 11 if and only if the alternating sum of its digits is divisible by 11. To get the alternating sum, add the first digit, subtract the second, add the third, and so on.
22. Let a be a number written in base 9. Show that a is divisible by 8 if and only if the sum of the base 9 digits is divisible by 8.

    (Recall that $a = (a_n a_{n-1} \cdots a_1 a_0)_9$ is the base 9 representation of a if $0 \leq a_i \leq 8$ for each i and $a = a_n \cdot 9^n + a_{n-1} \cdot 9^{n-1} + \cdots + a_1 \cdot 9 + a_0$. The sum of the base 9 digits is then $a_n + a_{n-1} + \cdots + a_1 + a_0$.)
23. The greatest common divisor of three integers is the largest integer dividing all three. If a, b, c are nonzero integers, prove that $\gcd(a, b, c) = \gcd(a, \gcd(b, c))$.
24. Let n be a positive integer and let $\overline{a} \in \mathbb{Z}_n$.

    a. Prove that the additive inverse of $\overline{a}$ is unique. In other words, prove that if $\overline{a} + \overline{x} = \overline{0} = \overline{a} + \overline{y}$, then $\overline{x} = \overline{y}$.
    b. Prove that the additive inverse of $\overline{a}$ is $\overline{-a}$. In other words, prove that $-(\overline{a}) = \overline{-a}$.
25. Solve the equation $\overline{14} \cdot x = \overline{3}$ in $\mathbb{Z}_{17}$ by first writing $1 = \gcd(14, 17)$ as a linear combination of 14 and 17.
26. The *Twin Primes Conjecture* is an open question asking whether or not there are infinitely many *twin primes*; a pair of twin primes is a pair of primes whose difference is 2. Prove that $(3, 5, 7)$ is the only set of *triplet primes*. That is, if $a, a + 2, a + 4$ are all prime numbers, prove that $a = 3$.

    (Hint: show that $a, a + 2, a + 4$ are distinct modulo 3).

1.3 Error Detection with Identification Numbers

The check digit provides no identification information, i.e., it is redundant. The purpose of this redundancy in an identification number is to detect errors in reading the number. We now discuss this idea in more detail and describe the three examples above in a unified way. First we need a version of the familiar dot product of vectors in $\mathbb{R}^3$; if $a = (a_1, \ldots, a_n)$ and $b = (b_1, \ldots, b_n)$ are n-tuples of numbers, set

$$a \cdot b = \sum_{i=1}^{n} a_i b_i = a_1 b_1 + \cdots + a_n b_n.$$

The test for validity in all three schemes is then represented in terms of the dot product. A ten digit number, or more accurately, a 10-tuple a of digits, is a valid zip code if $a \cdot (1,1,1,1,1,1,1,1,1,1)$ is divisible by 10. Likewise, a UPC is a 12-tuple a of digits such that $a \cdot (3,1,3,1,3,1,3,1,3,1,3,1)$ is divisible by 10. Finally, a 10-tuple a is a valid ISBN-10 provided that $a \cdot (10,9,8,7,6,5,4,3,2,1)$ is divisible by 11.

There are many more examples of identification number schemes, most constructed in the following manner: A modulus m and n-tuple $w = (w_1, \ldots, w_n)$ of positive integers are prescribed. A valid number for this scheme is an n-tuple $a = (a_1, \ldots, a_n)$ satisfying $a \cdot w \equiv 0 \bmod m$. For example, $w = (10,9,8,7,6,5,4,3,2,1)$ and $m = 11$ for the ISBN-10 scheme. Since the ith entry of w multiplies the ith entry of a in calculating the dot product, it gives a certain weight to that entry. For that reason w is referred to as the *weight vector* for the identification number scheme.

To be more precise, suppose we have an n-tuple w consisting of positive integers, and let m be a positive integer. We can make an identification number scheme as follows. Consider n-tuples $a = (a_1, \ldots, a_n)$ where $0 \le a_i \le m-1$. Then a is a valid identification number if $a \cdot w \equiv 0 \bmod m$. We will refer to this scheme as the identification number scheme associated with the weight vector w and with the integer m. The reason for the restriction on the a_i is that a_i and $a_i + m$ are the same modulo m. Therefore, the scheme cannot detect the difference between these two numbers. To eliminate this problem, we need to restrict the choices of the a_i so that two different possibilities for a_i are distinguished by the scheme; that is, any two possibilities must be distinct modulo m.

The errors which occur most frequently in identification number schemes are single digit errors and transposition errors, in which adjacent digits are switched. The next proposition shows how to design a scheme which will detect every single digit error.

Proposition 1.25. *Given w and m as above. Then every error in reading the ith entry of a can be detected provided that each w_i is relatively prime to m.*

Proof. Let $a = (a_1, \ldots, a_n)$ be a valid identification number and that only the ith entry a_i is changed to another number, say $b_i \neq a_i$, with $0 \le b_i \le m-1$. If $w = (w_1, \ldots, w_n)$, then, because a is a valid number, we have in $\mathbb{Z}_m$

$$\overline{0} = \overline{\sum_{i=1}^{n} a_i w_i} = \sum_{i=1}^{n} \overline{a_i w_i},$$

where the last equality comes from the definition of the operations in $\mathbb{Z}_m$. If $b = (a_1, \ldots, a_{i-1}, b_i, a_{i+1}, \ldots, a_n)$ is the number obtained by replacing the ith entry of a to b_i, set $\epsilon = b_i - a_i \neq 0$ so that $b_i = a_i + \epsilon$. Note that $0 \le a_i, b_i \le m-1$ so that $0 < \epsilon \le m-1$. The validation computation for b becomes

$$\begin{aligned}\overline{b \cdot w} &= \sum_{j=1}^{i-1} \overline{a_j w_j} + \overline{b_i}\overline{w_i} + \sum_{j=i+1}^{n} \overline{a_j w_j} \\ &= \sum_{j=1}^{i-1} \overline{a_j w_j} + \overline{(a_i + \epsilon)}\overline{w_i} + \sum_{j=i+1}^{n} \overline{a_j w_j} \\ &= \overline{a \cdot w} + \overline{\epsilon w_i}.\end{aligned}$$

But a is a valid word, so $\overline{a \cdot w} = \overline{0}$ and therefore $\overline{b \cdot w} = \overline{\epsilon w_i}$.

Suppose that the error is undetected, so that $\overline{0} = \overline{b \cdot w} = \overline{\epsilon w_i}$. Assuming that w_i is relatively prime to m, Corollary 1.17 tells us that there is an $\overline{x}$ with $\overline{w_i x} = \overline{1}$ in $\mathbb{Z}_m$. Multiplying both sides of the equation $\overline{0} = \overline{\epsilon w_i}$ by $\overline{x}$ and simplifying gives $\overline{0} = \overline{\epsilon} = \overline{b_i} - \overline{a_i}$, or $\overline{b_i} = \overline{a_i}$. However, this forces $b_i = a_i$ since both a_i and b_i are between 0 and $m - 1$. This is a contradiction, so $b \cdot w$ cannot be zero, and hence the error $a_i \mapsto b_i$ in reading the ith entry of a will be detected by this identification number scheme. □

Corollary 1.26. *If each entry of w is relatively prime to m, then the identification number scheme associated with w and m can detect a single error in any digit.*

We now return to the three schemes we discussed earlier. The zip code scheme uses the weight vector $w = (1, 1, 1, 1, 1, 1, 1, 1, 1, 1)$ and $m = 10$. Each entry of w is clearly relatively prime to 10, so the zip code scheme will detect all single errors. The UPC scheme uses $w = (3, 1, 3, 1, 3, 1, 3, 1, 3, 1, 3, 1)$ and $m = 10$. Both 1 and 3 are relatively prime to 10, so the UPC scheme also detects all single errors. Clearly an identification number scheme with $m = 10$, and weight vector w having entries from the set $\{1, 3, 7, 9\}$, will detect all single errors. The ISBN-10 scheme, with $m = 11$ and weight vector $w = (10, 9, 8, 7, 6, 5, 4, 3, 2, 1)$, detects all single errors since every entry of w is certainly relatively prime to the prime number 11.

The corollary above tells us when an identification number scheme detects a single error. No scheme of the type we have discussed will always detect errors in more than one digit. For example, if the zip code 8800380012 is replaced by 7900380012 by changing the first two digits, then this number satisfies the test for validity. Similarly, if the UPC 0 49000 01134 0 is changed to 0 49000 01163 0 by changing the 10th and 11th digits, as shown, the number is still valid.

Now let's consider transposition errors since a tendency to transpose digits is common when reading numbers. For example, the zip code 8800380012 might be read as 8800830012, by interchanging the 3 and an 8. Some coding schemes detect this type of error, and some do not. For example, the zip code scheme does not detect interchanging of digits. This is because the validation test here is simply that the sum of the ten digits must be divisible by 10 and the commutative law for addition says that interchanging two digits does not change the sum. UPC can detect some transposition errors, e.g. interchanging an adjacent $(1, 2)$ pair, but not all, e.g. interchanging an adjacent $(1, 6)$. However, as is to be seen in a homework exercise, the ISBN-10 scheme detects every interchanging of digits. For example, with the ISBN-10 0387947531, interchanging the second and third digits yields 0837947531. The test for validity becomes

$$(0, 8, 3, 7, 9, 4, 7, 5, 3, 1) \cdot (10, 9, 8, 7, 6, 5, 4, 3, 2, 1) = 269,$$

which is not divisible by 11. Therefore, the number 0837947531 is not valid. Similarly, if we interchange the final two digits, we get 0387947513, and

$$(0, 3, 8, 7, 9, 4, 7, 5, 1, 3) \cdot (10, 9, 8, 7, 6, 5, 4, 3, 2, 1) = 262,$$

which is also not divisible by 11. A special case of the sensitivity of ISBN-10 to transposition errors is proved here with a much more general case left to Exercise 5.

Proposition 1.27. *Suppose that* $(a_1, a_2, \ldots, a_{10})$ *is a valid ISBN-10. If* $a_1 \neq a_2$*, then* $(a_2, a_1, a_3, \ldots, a_{10})$ *is not a valid ISBN-10. Thus, the ISBN-10 scheme detects transposition of the first two digits.*

Proof. Let $a = (a_1, a_2, \ldots, a_{10})$ be a valid ISBN-10. If

$$w = (10, 9, 8, 7, 6, 5, 4, 3, 2, 1),$$

then $\overline{a \cdot w} = \overline{0}$ in $\mathbb{Z}_{11}$. Suppose that $a_1 \neq a_2$, and set $b = (a_2, a_1, \ldots, a_{10})$. Then

$$\begin{aligned} \overline{a \cdot w - b \cdot w} &= \overline{10a_1 + 9a_2 - 10a_2 - 9a_1} \\ &= \overline{a_1 - a_2}. \end{aligned}$$

Since $a_1 \neq a_2$ and $0 \leq a_1, a_2 \leq 10$, we see that $\overline{a_1} \neq \overline{a_2}$. Consequently, the equation above shows that $\overline{a \cdot w - b \cdot w} \neq \overline{0}$, or $\overline{b \cdot w} \neq \overline{a \cdot w} = \overline{0}$. Thus, b is not a valid ISBN-10. □

Exercises

1. Let $(a_1, \ldots, a_{10})$ be an invalid ISBN-10. Show that any single digit of **a** can be changed appropriately to give a valid ISBN.
2. Let $(a_1, \ldots, a_{12})$ be a valid UPC. Show that the error in transposing the first two digits of this number is detected by the UPC scheme if and only if $a_2 - a_1$ is not divisible by 5. Use this to give an example of a UPC $(a_1, \ldots, a_{12})$ with $a_1 \neq a_2$ but such that $(a_2, a_1, a_3 \ldots, a_{12})$ is also a valid UPC.
3. Consider the following identification number scheme: if $w = (3, 5, 2, 7)$, then a 4-tuple $a = (a_1, a_2, a_3, a_4)$, with each a_i an integer with $0 \leq a_i \leq 7$, is valid if and only if $\overline{a \cdot w} = \overline{0}$ in $\mathbb{Z}_8$. Show that this scheme detects any error in the first, second, and fourth digit. Give an example of a valid codeword and an error in the third digit of the codeword that is not detected.
4. Define an identification number scheme as follows: set $w = (2, 5, 6, 4, 7)$, and a 5-tuple $a = (a_1, a_2, a_3, a_4, a_5)$ is a valid number if $0 \leq a_i \leq 8$ for each i, and if $a \cdot w$ is divisible by 9. Determine which single errors are detected by this scheme. That is, determine for which i an error in reading the ith digit is always detected. Describe how you can change w in order to guarantee that an error in any digit is always detected.
5. Prove that transposition of any two digits can be detected with the ISBN-10 scheme. That is, if $(a_1, \ldots, a_{10})$ is a valid ISBN-10, and if $i < j$ with $a_i \neq a_j$, show that $(a_1, \ldots, a_j, \ldots, a_i, \ldots, a_{10})$ is not a valid ISBN-10.
6. The number 0-387-79847-X was obtained from a valid ISBN-10 by transposing two digits. Can you tell which two digits were transposed? Either explain how you can tell which digits were transposed or give an example of two valid ISBN-10s obtained by transposing two digits of this number.
7. Construct a length 5 check digit scheme that detects every single digit error and every transposition error.
8. A jump transposition error is one in which the valid codeword

$$(a_1, \ldots, a_{i-1}, a_i, a_{i+1}, \ldots, a_n)$$

is misread as

$$(a_1, \ldots, a_{i+1}, a_i, a_{i-1}, \ldots, a_n).$$

Exercise 5 shows that ISBN-10 detects every jump transposition error. Show that ISBN-13 doesn't detect any.

References

1. Dudley U (2008) Elementary number theory, 2nd edn. Dover Books, New York
2. Johnson B, Richman F (1997) Numbers and symmetry: an introduction to algebra. CRC, Boca Raton

Chapter 2
Error Correcting Codes

The identification number schemes we discussed in the previous chapter give us the ability to determine if an error has been made in recording or transmitting information. However, they are limited in two ways. First, the types of errors detected are fairly restrictive, e.g. single digit errors or interchanging digits. Second, they provide no way to recover the intended information. Some more sophisticated ideas and mathematical concepts enable methods to encoding and transmit information in ways that allow both detection and correction of errors. There are many applications of these so-called error correcting codes, among them transmission of digital images from planetary probes and playing compact discs and DVD movies.

2.1 Basic Notions

To discuss error correcting codes, we need first to set the context and define some terms. We work throughout in binary; that is, we will work over $\mathbb{Z}_2$. To simplify notation, we will write the two elements of $\mathbb{Z}_2$ as 0 and 1 instead of as $\overline{0}$ and $\overline{1}$. If n is a positive integer, then the set $\mathbb{Z}_2^n$ is the set of all n-tuples of $\mathbb{Z}_2$-entries. Elements of $\mathbb{Z}_2^n$ are called *words*, or words of length n. For convenience we will write elements of $\mathbb{Z}_2^n$ either with the usual notation, or as a concatenation of digits. For instance, we will write $(0, 1, 0, 1)$ and 0101 for the same 4-tuple. We can equip $\mathbb{Z}_2^n$ with an operation of addition by using point-wise addition. That is, we define

$$(a_1, \ldots, a_n) + (b_1, \ldots, b_n) = (a_1 + b_1, \ldots, a_n + b_n).$$

A consequence of the fact that $0 + 0 = 0 = 1 + 1$ in $\mathbb{Z}_2$ is that $a + a = \mathbf{0}$ for every $a \in \mathbb{Z}_2^n$, where $\mathbf{0}$ is the vector $(0, \ldots, 0)$ consisting of all zeros.

A *linear code* of length n is a nonempty subset of $\mathbb{Z}_2^n$ that is closed under the addition in $\mathbb{Z}_2^n$. Although nonlinear codes exist and are studied, linear codes are used most frequently in applications and much of the discussion simplifies greatly in this context. Because of their importance, we will consider only linear codes and drop the adjective "linear" from now on. We will refer to elements of a code as codewords.

Electronic supplementary material The online version of this chapter (doi:10.1007/978-3-319-04498-9_2) contains supplementary material, which is available to authorized users. The supplementary material can also be downloaded from http://extras.springer.com.

D.R. Finston and P.J. Morandi, *Abstract Algebra: Structure and Application*,
Springer Undergraduate Texts in Mathematics and Technology, DOI 10.1007/978-3-319-04498-9_2

Example 2.1. The set $\{00, 01, 10, 11\} = \mathbb{Z}_2^2$ is a code of length 2, and the set $\{0000, 1010, 0101, 1111\}$, which is a proper subset of $\mathbb{Z}_2^4$, is a code of length 4.

Let $w = a_1 \cdots a_n$ be a word of length n. Then the *weight* of w is the number of digits of w equal to 1. We denote the weight of w by $\mathrm{wt}(w)$. An equivalent and useful way to think about the weight of the word $w = a_1 \cdots a_n$ is to treat the a_i as the integers 0 or 1 (rather than as residue classes for the moment) and note that

$$wt(w) = \sum_{i=1}^{n} a_i .$$

There are some obvious consequences of this definition. First of all, $\mathrm{wt}(w) = 0$ if and only if $w = \mathbf{0}$. Second, $\mathrm{wt}(w)$ is a nonnegative integer. A more sophisticated fact about weight is its relation with addition. If $v, w \in \mathbb{Z}_2^n$, then $\mathrm{wt}(v+w) \leq \mathrm{wt}(v) + \mathrm{wt}(w)$. This is true because cancellation occurs when the ith components of v and w are both equal to 1. More precisely, write x_i for the ith component of a word x. The weight of x is then given by the equation $\mathrm{wt}(x) = |\{i : 1 \leq i \leq n, x_i = 1\}|$. Note that $(v + w)_i = v_i + w_i$, so that $(v + w)_i = 1$ implies that either $v_i = 1$ or $w_i = 1$ (but not both). Therefore,

$$\{i : 1 \leq i \leq n, (v + w)_i = 1\} \subseteq \{i : v_i = 1\} \cup \{i : w_i = 1\} .$$

Since $|A \cup B| \leq |A| + |B|$ for any two finite sets A, B, the inclusion above and the latter description of weight yields $\mathrm{wt}(v + w) \leq \mathrm{wt}(v) + \mathrm{wt}(w)$, as desired.

The idea of weight gives a notion of distance on $\mathbb{Z}_2^n$. If v, w are words, then we set the *distance* $D(v, w)$ between v and w to be

$$D(v, w) = \mathrm{wt}(v + w).$$

Alternatively, $D(v, w)$ is equal to the number of positions in which v and w differ. The function D shares the basic properties of distance in Euclidean space $\mathbb{R}^3$. More precisely, it satisfies the properties of the following lemma.

Lemma 2.2. *The distance function D defined on $\mathbb{Z}_2^n \times \mathbb{Z}_2^n$ satisfies:*

1. $D(v, v) = 0$ for all $v \in \mathbb{Z}_2^n$;
2. For any $v, w \in \mathbb{Z}_2^n$, if $D(v, w) = 0$, then $v = w$;
3. $D(v, w) = D(w, v)$ for any $v, w \in \mathbb{Z}_2^n$;
4. The Triangle Inequality: $D(v, w) \leq D(v, u) + D(u, w)$ for any $u, v, w \in \mathbb{Z}_2^n$.

Proof. Since $v + v = \mathbf{0}$, we have $D(v, v) = \mathrm{wt}(v + v) = \mathrm{wt}(\mathbf{0}) = 0$. This proves (1). We note that $\mathbf{0}$ is the only word of weight 0. Thus, if $D(v, w) = 0$, then $\mathrm{wt}(v + w) = 0$, which forces $v + w = \mathbf{0}$. However, adding w to both sides yields $v = w$, and this proves (2). The equality $D(v, w) = D(w, v)$ is obvious since $v + w = w + v$. Finally, we prove (4), the only non-obvious statement, with a cute argument. Given $u, v, w \in \mathbb{Z}_2^n$, we have, from the definition and the fact about the weight of a sum given above,

$$\begin{aligned} D(v, w) &= \mathrm{wt}(v + w) = \mathrm{wt}((v + u) + (u + w)) \\ &\leq \mathrm{wt}(v + u) + \mathrm{wt}(u + w) \\ &= D(v, u) + D(u, w). \end{aligned}$$

□

To discuss error correction we must first formalize the notion. Let C be a code. If w is a word, to correct, or decode, w means to select the codeword $v \in C$ such that

$$D(v, w) = \min\{D(u, w) : u \in C\}.$$

In other words, we decode w by choosing the closest codeword to w, under our notion of distance. There need not be a unique closest codeword, however. When this happens we can either randomly select a closest codeword, or do nothing. We refer to this notion of decoding as *maximum likelihood detection*, or MLD, the assumption being that the means of transmission of information is reliable so that if an error is introduced, the correct information is most likely to be the codeword that differs from the received word in the fewest number of positions.

Example 2.3. Let $C = \{00000, 10000, 011000, 11100\}$. If $w = 10001$, then w is distance 1 from 10000 and distance more than 1 from the other two codewords. Thus, we would decode w as 10000. However, if $u = 11000$, then u is distance 1 from both 10000 and from 111000. Thus, either is an appropriate choice to decode u.

We now define what it means for a code to be an error correcting code.

Definition 2.4. Let C be a code and let t be a positive integer. Then C is a t-error correcting code if whenever a word w differs from the nearest codeword v by a distance of at most t, then v is the unique closest codeword to w.

If a codeword v is transmitted and received as w, we can express w as $v + u$, and we say that $u = v + w$ is the error in transmission. As a word, the error u has a certain weight. So C is t-error correcting if for every codeword v and every word u whose weight is at most t, then v is the unique closest codeword to $v + u$.

If C is a t-error correcting code, then we say that C corrects t errors. Thus one way of interpreting the definition is that if v is a codeword, and if w is obtained from v by changing at most t entries of v, then v is the unique closest codeword to w. Therefore, by MLD decoding, w will be decoded as v.

Example 2.5. The code $C = \{000000, 111000, 000111\}$ is 1-error correcting. A word which differs from 000000 in one entry differs from the other two codewords in at least two entries. Similarly for the other two codewords in C.

Example 2.6. The code $C = \{00000, 10000, 011000, 11100\}$ above corrects no errors. Note that the word $u = 11000$ given in that example is a distance 1 from a codeword, but that codeword is not the unique closest codeword to u.

To determine for which t a code corrects t errors, we relate error correction to the distance of a code.

Definition 2.7. The distance d of a code is defined by

$$d = \min\{D(u, v) : u, v \in C, u \neq v\}.$$

For intuitive purposes it may be useful to think of the minimum distance as the diameter of the smallest circle containing at least two codewords.

We denote by $\lfloor a \rfloor$ the greatest integer less than or equal to the number a.

Proposition 2.8. *Let C be a code of distance d and set $t = \lfloor (d-1)/2 \rfloor$. Then C is a t-error correcting code but not a $(t+1)$-error correcting code.*

Proof. To prove that C is t-error correcting, let w be a word, and suppose that v is a codeword with $D(v, w) \leq t$. We need to prove that v is the unique closest codeword to w. We do this by proving that $D(u, w) > t$ for any codeword $u \neq v$. If not, suppose that u is a codeword with $u \neq v$ and $D(u, w) \leq t$. Then, by the Triangle Inequality,

$$D(u, v) \leq D(u, w) + D(w, v) \leq t + t = 2t < d.$$

This is a contradiction to the definition of d. Thus v is indeed the unique closest codeword to w.

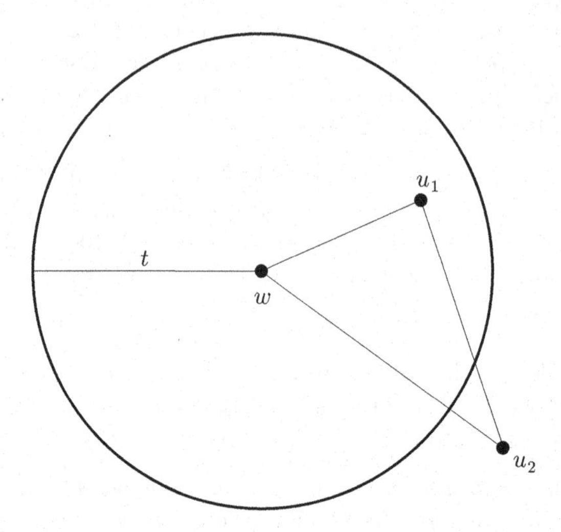

To finish the proof, we need to prove that C does not correct $t + 1$ errors. Since the code has distance d, there are codewords u_1, u_2 with $d = D(u_1, u_2)$; in other words, u_1 and u_2 differ in exactly d positions. Let w be the word obtained from u_1 by changing exactly $t + 1$ of those d positions. Then $D(u_1, w) = t + 1$ and $D(u_2, w) = d - (t + 1)$. Since $t = \lfloor (d - 1)/2 \rfloor$ by our assumption, $(d - 2)/2 \leq t \leq (d - 1)/2$. In particular, $d - 2 \leq 2t$ so that $D(u_2, w) = d - (t + 1) \leq t + 1$. Thus u_1 is not the unique closest codeword to w, since u_2 is either equally close or closer to w. Therefore C is not a $(t + 1)$-error correcting code.

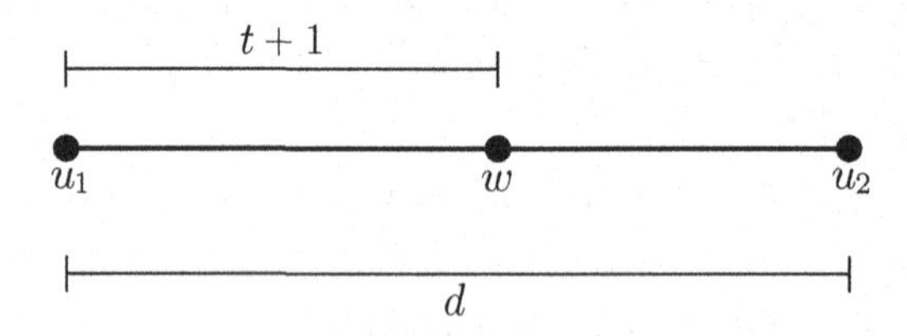

□

Example 2.9. Let $C = \{00000, 00111, 11100, 11011\}$. The distance of C is 3, and so C is a 1-error correcting code.

Example 2.10. Let n be an odd positive integer, and let $C = \{0 \cdots 0, 1 \cdots 1\}$ be a code of length n. If $n = 2t + 1$, then C is a t-error correcting code since the distance of C is n. Thus, by making the length of C long enough, we can correct any number of errors that we wish. However, note that the fraction of components of a word that can be corrected is t/n, and this is always less than $1/2$.

Exercises

1. Find distance and error correction capability of the following codes:

   (a) $\{0000000, 1010101, 0101010, 1111111\}$,

   (b) $\{00000000, 11111111, 11100000, 00011111\}$,

   (c) $\{00000000,\ 11110000,\ 00001111,\ 10101010,\ 11111111,\ 01011010,\ 10100101,\ 01010101\}$.

2. Construct a linear code of length 5 with more than two codewords that corrects one error. Can you construct a linear code of length 4 with more than two words that corrects one error?
3. Let C be the code consisting of the solutions to the matrix equation $Ax = \mathbf{0}$, where

$$A = \begin{pmatrix} 1\,0\,1\,1\,1\,0 \\ 0\,1\,1\,1\,0\,1 \\ 1\,1\,1\,0\,0\,0 \end{pmatrix}.$$

   Determine the codewords of C, and determine the distance and error correction capability of C.
4. Let A be a matrix, and let C be the code consisting of all solutions to $Ax = \mathbf{0}$. If A has neither a column of zeros nor two equal columns, prove that the distance of C is at least 3.
   (Hint: If v has weight 1 or weight 2, look at how Av can be written in terms of the columns of A.)
5. Let C be a code such that if $u, v \in C$, then $u + v \in C$. Prove that the distance of C is equal to the smallest weight of a nonzero codeword.
6. Let C be the code consisting of all solutions to a matrix equation $Ax = \mathbf{0}$. Let d be the largest integer such that any sum of fewer than d columns of A is nonzero. Prove that C has distance d.

2.2 Gaussian Elimination

In this section we recall some basic results about matrices, in particular Gaussian elimination, rank, and nullity. Our immediate concern is with matrices whose entries lie in $\mathbb{Z}_2$ in order to discuss the Hamming and Golay codes, historically the first examples of error correcting codes.

A system of linear equations is equivalent to a single matrix equation $AX = b$, where A is the matrix of coefficients, and X is the column matrix of variables. For example, the system of linear equations over the rational numbers

$$2x + 3y - z = 1$$
$$x - y + 5z = 2$$

is equivalent to the matrix equation

$$\begin{pmatrix} 2 & 3 & -1 \\ 1 & -1 & 5 \end{pmatrix} \begin{pmatrix} x \\ y \\ z \end{pmatrix} = \begin{pmatrix} 1 \\ 2 \end{pmatrix}.$$

The primary matrix-theoretic method for solving such a system is Gaussian elimination on the augmented matrix obtained from the coefficient matrix by appending on its right the column consisting of the right-hand side of the equation. Recall that Gaussian elimination employs operations on the rows of a matrix, with the end result a matrix in row reduced echelon form. The latter represents a system of equations whose solutions, which are identical to those of the original system, can be found easily.

The three elementary row operations are :

- Replacing a row with a multiple of it by a nonzero scalar,
- Interchanging two rows,
- Replacing a row by its sum with a scalar multiple of another row.

In $\mathbb{Z}_2$ arithmetic the only multipliers available are 0 and 1 and $1 + 1 = 0$ in $\mathbb{Z}_2$ (so that $1 = -1$ and subtraction is the same operation as addition). In this context, the first of the three row operations listed above is not useful, since multiplying a row by 1 does not affect the row, and the third operation reduces to adding one row to another. The desired outcome is a matrix in row reduced echelon form:

Definition 2.11. A matrix A is in row reduced echelon form if all three of the following conditions are satisfied:

1. The first nonzero entry of each row is 1. This entry is called a leading 1.
2. If a column contains a leading 1, then all other entries of the column are 0.
3. If $i > j$, and if row i and row j each contain a leading 1, then the column containing the leading 1 of row i is further to the right than the column containing the leading 1 of row j.

To help understand Condition 3 of the definition, the leading 1's go to the right as you go from top to bottom in the matrix, so that the matrix is in some sense triangular.

Example 2.12. The following matrices over $\mathbb{Z}_2$ are in row reduced echelon form:

$$\begin{pmatrix} 1 & 1 \\ 0 & 0 \end{pmatrix} \quad \begin{pmatrix} 1 & 0 & 1 & 0 \\ 0 & 1 & 1 & 0 \\ 0 & 0 & 0 & 1 \end{pmatrix} \quad \begin{pmatrix} 1 & 0 & 0 \\ 0 & 1 & 0 \\ 0 & 0 & 1 \\ 0 & 0 & 0 \end{pmatrix}.$$

The columns with leading ones have the form of vectors e_i with a 1 in the ith position and 0's elsewhere.

In Chap. 4, familiar concepts from linear algebra over the real numbers will be systematically extended to include linear algebra over $\mathbb{Z}_2$. For now though, let's recall some facts about matrices with real entries in $\mathbb{R}$ that also hold for matrices with entries in $\mathbb{Z}_2$. First, the *row space* of a matrix is the vector space spanned by its rows. If the matrix is $m \times n$, then the rows are n-tuples, so the row space is a subspace of the space of all n-tuples. Since Gaussian elimination operates on the rows of a matrix in a reversible way, the row space of a matrix is identical with that of its row reduced echelon form. The *column space* of a matrix is the space spanned by the columns of the matrix. Again, if the matrix is $m \times n$, then the columns are m-tuples, so the column space is a subspace of the space of all m-tuples. These observations hold as well for matrices with entries in $\mathbb{Z}_2$. The only difference is that the span of a collection of rows or columns is merely the sum of some subset of them, again because the only multipliers available are 0 and 1.

The dimension of a vector space over $\mathbb{R}$ is the number of elements in a basis, provided this is finite. Otherwise the dimension is infinite. For an $m \times n$ matrix A, the dimension of the row space and the dimension of the column space are always finite and equal; this integer is called the *rank* of A. One benefit to reducing A to its row reduced echelon form E_A is that the nonzero rows of E_A (i.e., those that contain a leading 1) form a basis for the row space of A. Consequently, the dimension of the row space is the number of nonzero rows in E_A. Thus, an alternative definition of the rank of a matrix is the number of leading 1's in the row reduced echelon form obtained from the matrix. Again these assertions hold for matrices with entries in $\mathbb{Z}_2$.

The fact that the homogeneous linear systems $AX = 0$ and $E_A X = 0$ have the same solutions can be interpreted as the statement that the columns of A and the columns of E_A have the identical dependence relations (but their column spaces may be different). From Condition 2 it is clear that the columns of E_A that contain the leading 1's form a basis for its column space. Call these columns $c_{i1}, \dots c_{ir}$. But then columns $i_1, \dots i_r$ of the matrix A form a basis for its column space, hence the assertion above about the equality of the "row rank" and "column rank." It is clear also that the maximum possible rank of an $m \times n$ matrix is the minimum of m and n (although the matrix $\begin{pmatrix} 1 & 0 \\ 0 & 0 \end{pmatrix}$, for instance, shows that this bound need not be achieved).

Even though you might be most familiar with matrices whose entries are real numbers, the row operations above require only the ability to add, subtract, multiply, and divide the entries. In many situations, matrices arise whose entries are not real numbers, and our initial work in coding theory leads to matrices whose entries lie in $\mathbb{Z}_2$ (wherein we can certainly add, subtract, multiply, and divide, with the usual proscription against division by 0). Furthermore, all the theorems of linear algebra have analogues to this setting, and later on the fundamentals of linear algebra will be generalized to include other sets of scalars. Again, all that is necessary is closure of the scalars under the four arithmetic operations and the standard arithmetic properties analogous to those that hold for real number arithmetic (i.e., commutativity and associativity of addition and multiplication, and distributivity of multiplication over addition).

We now give several examples of reducing matrices with $\mathbb{Z}_2$ entries to echelon form. In each example once we have the matrix in row reduced echelon form, the leading 1's are marked in boldface.

Example 2.13. Consider the matrix

$$A = \begin{pmatrix} 1 & 0 & 0 & 1 \\ 1 & 1 & 0 & 1 \\ 0 & 1 & 1 & 1 \end{pmatrix}.$$

We reduce the matrix with the following steps. You should determine which row operation was done in each step.

$$\begin{pmatrix} 1 & 0 & 0 & 1 \\ 1 & 1 & 0 & 1 \\ 0 & 1 & 1 & 1 \end{pmatrix} \Longrightarrow \begin{pmatrix} 1 & 0 & 0 & 1 \\ 0 & 1 & 0 & 0 \\ 0 & 1 & 1 & 1 \end{pmatrix} \Longrightarrow \begin{pmatrix} \mathbf{1} & 0 & 0 & 1 \\ 0 & \mathbf{1} & 0 & 0 \\ 0 & 0 & \mathbf{1} & 1 \end{pmatrix}.$$

The rank of A is equal to 3.

Example 2.14. Consider the matrix

$$A = \begin{pmatrix} 1 & 1 & 0 & 0 & 1 & 0 \\ 1 & 0 & 1 & 0 & 0 & 1 \\ 0 & 1 & 1 & 1 & 1 & 0 \\ 0 & 0 & 0 & 1 & 0 & 1 \end{pmatrix}.$$

To reduce this matrix, we can do the following steps.

$$\begin{pmatrix} 1\,1\,0\,0\,1\,0 \\ 1\,0\,1\,0\,0\,1 \\ 0\,1\,1\,1\,1\,0 \\ 0\,1\,1\,0\,1\,1 \end{pmatrix} \Longrightarrow \begin{pmatrix} 1\,1\,0\,0\,1\,0 \\ 0\,1\,1\,0\,1\,1 \\ 0\,1\,1\,1\,1\,0 \\ 0\,1\,1\,0\,1\,1 \end{pmatrix} \Longrightarrow \begin{pmatrix} 1\,0\,1\,0\,0\,1 \\ 0\,1\,1\,0\,1\,1 \\ 0\,1\,1\,1\,1\,0 \\ 0\,1\,1\,0\,1\,1 \end{pmatrix}$$

$$\Longrightarrow \begin{pmatrix} 1\,0\,1\,0\,0\,1 \\ 0\,1\,1\,0\,1\,1 \\ 0\,0\,0\,1\,0\,1 \\ 0\,1\,1\,0\,1\,1 \end{pmatrix} \Longrightarrow \begin{pmatrix} \mathbf{1}\,0\,1\,0\,0\,1 \\ 0\,\mathbf{1}\,1\,0\,1\,1 \\ 0\,0\,0\,\mathbf{1}\,0\,1 \\ 0\,0\,0\,0\,0\,0 \end{pmatrix}.$$

The rank of A is equal to 3.

We now illustrate how the row reduced echelon form yields the solution of the systems of equations giving rise to the matrices in the previous examples.

Example 2.15. The system of equations

$$\begin{aligned} x &= 1 \\ x + y &= 1 \\ y + z &= 1 \end{aligned}$$

has augmented matrix

$$\begin{pmatrix} 1\,0\,0\,1 \\ 1\,1\,0\,1 \\ 0\,1\,1\,1 \end{pmatrix}.$$

The reduction of this matrix

$$\begin{pmatrix} 1\,0\,0\,1 \\ 0\,1\,0\,0 \\ 0\,0\,1\,1 \end{pmatrix}.$$

corresponds to the system of equations

$$\begin{aligned} x &= 1 \\ y &= 0 \\ z &= 1 \end{aligned}$$

and hence solves the original system.

Example 2.16. The augmented matrix

$$\begin{pmatrix} 1\,1\,0\,0\,1\,0 \\ 1\,0\,1\,0\,0\,1 \\ 0\,1\,1\,1\,1\,0 \\ 0\,1\,1\,0\,1\,1 \end{pmatrix}$$

corresponds to the system of equations

$$\begin{aligned} x_1 + x_2 + x_5 &= 0 \\ x_1 + x_3 &= 1 \\ x_2 + x_3 + x_4 + x_5 &= 0 \\ x_2 + x_3 + x_5 &= 1. \end{aligned}$$

Reducing the matrix yields

$$\begin{pmatrix} \mathbf{1}\,0\,1\,0\,0\,1 \\ 0\,\mathbf{1}\,1\,0\,1\,1 \\ 0\,0\,0\,\mathbf{1}\,0\,1 \\ 0\,0\,0\,0\,0\,0 \end{pmatrix},$$

which corresponds to the system of equations

$$\begin{aligned} x_1 + x_3 &= 1 \\ x_2 + x_3 + x_5 &= 1 \\ x_4 &= 1. \end{aligned}$$

The leading 1's in boldface in the echelon matrix correspond to the variables x_1, x_2, and x_4. Solving for these yields the full solution

$$\begin{aligned} x_1 &= 1 + x_3, \\ x_2 &= 1 + x_3 + x_5 \\ x_4 &= 1 \\ & x_3 \text{ and } x_5 \text{ are arbitrary.} \end{aligned}$$

We can write out all solutions to this system of equations, since each of x_3 and x_5 can take on the two values 0 and 1. This gives us four solutions, which we write as row vectors:

$$(x_1, x_2, x_3, x_4, x_5) = (1 + x_3, 1 + x_3 + x_5, x_3, 1, x_5),$$

where $x_3 \in \{0, 1\}$ and $x_5 \in \{0, 1\}$.

The general solution is

$$(1 + x_3, 1 + x_3 + x_5, x_3, 1, x_5) = (1, 1, 0, 1, 0) + x_3(1, 1, 1, 0, 0) + x_5(0, 1, 0, 0, 1)$$

so that $(1, 1, 0, 1, 0)$, which corresponds to the values $x_3 = x_5 = 0$, yields a particular solution to the linear system. On the other hand, the vectors $(1, 1, 1, 0, 0), (0, 1, 0, 0, 1)$ solve the homogeneous system

$$x_1 + x_2 + x_5 = 0,$$
$$x_1 + x_3 = 0,$$
$$x_2 + x_3 + x_4 + x_5 = 0,$$
$$x_2 + x_3 + x_5 = 0.$$

(Check this!) Thus any solution to the inhomogeneous system is obtained as the sum of a particular solution and a solution to the associated homogenous system.

Example 2.17. Let H be the *Hamming matrix* (named for Richard Hamming, mathematician, pioneer computer scientist, and inventor of the Hamming error correcting codes):

$$H = \begin{pmatrix} 0\,0\,0\,1\,1\,1\,1 \\ 0\,1\,1\,0\,0\,1\,1 \\ 1\,0\,1\,0\,1\,0\,1 \end{pmatrix},$$

and consider the homogeneous system of equations $HX = \mathbf{0}$, where $\mathbf{0}$ refers to the 3×1 zero matrix and X is a 7×1 matrix of the variables $x_1, \ldots, x_7$. To solve this system we reduce the augmented matrix in one step to

$$\begin{pmatrix} 0\,0\,0\,1\,1\,1\,1\,0 \\ 0\,1\,1\,0\,0\,1\,1\,0 \\ 1\,0\,1\,0\,1\,0\,1\,0 \end{pmatrix},$$

yielding

$$\begin{pmatrix} \mathbf{1}\,0\,1\,0\,1\,0\,1\,0 \\ 0\,\mathbf{1}\,1\,0\,0\,1\,1\,0 \\ 0\,0\,0\,\mathbf{1}\,1\,1\,1\,0 \end{pmatrix}.$$

This matrix corresponds to the system of equations

$$x_1 + x_3 + x_5 + x_7 = 0,$$
$$x_2 + x_3 + x_6 + x_7 = 0,$$
$$x_4 + x_5 + x_6 + x_7 = 0.$$

Again, we have marked the leading 1's in boldface, and the corresponding variables can be solved in terms of the others, which can be arbitrary. So, the solution to this system is

$$x_1 = x_3 + x_5 + x_7,$$
$$x_2 = x_3 + x_6 + x_7,$$
$$x_4 = x_5 + x_6 + x_7,$$
$$x_3, x_5, x_6, x_7 \text{ are arbitrary.}$$

Since we have four variables, x_3, x_5, x_6, and x_7, that can take on the values 0 or 1 in $\mathbb{Z}_2$ arbitrarily, there are exactly $2^4 = 16$ solutions to this system of equations.

To finish this chapter, we recall a theorem that will help us determine numeric data about error correcting codes. Before stating the theorem we explore the context in which it will be applied and recall some terminology.

The *kernel*, or *nullspace*, of a matrix A is the set of all solutions to the homogeneous equation $AX = \mathbf{0}$. As an illustration, consider the Hamming matrix H of the previous example.

Example 2.18. The solution above to the homogeneous equation $HX = \mathbf{0}$ can be described systematically by determining a basis for the nullspace of H. Since each distinct choice of the variables x_3, x_5, x_6, and x_7 in $\mathbb{Z}_2$ results in a unique solution to $HX = \mathbf{0}$, we obtain 4 solutions by successively setting one of these variables equal to 1 and all others arbitrary variables equal to 0, then using

$$\begin{aligned} x_1 &= x_3 + x_5 + x_7, \\ x_2 &= x_3 + x_6 + x_7, \\ x_4 &= x_5 + x_6 + x_7 \end{aligned}$$

to determine the values for the remaining variables. This technique results in the vectors

$$\begin{pmatrix} 1 \\ 1 \\ 1 \\ 0 \\ 0 \\ 0 \\ 0 \end{pmatrix}, \begin{pmatrix} 1 \\ 0 \\ 0 \\ 1 \\ 1 \\ 0 \\ 0 \end{pmatrix}, \begin{pmatrix} 0 \\ 1 \\ 0 \\ 1 \\ 0 \\ 1 \\ 0 \end{pmatrix}, \begin{pmatrix} 1 \\ 1 \\ 0 \\ 1 \\ 0 \\ 0 \\ 1 \end{pmatrix}$$

which form a basis for the nullspace of H. Indeed, the general solution of $HX = \mathbf{0}$ is given by

$$\begin{pmatrix} x_1 \\ x_2 \\ x_3 \\ x_4 \\ x_5 \\ x_6 \\ x_7 \end{pmatrix} = \begin{pmatrix} x_3 + x_5 + x_7 \\ x_3 + x_6 + x_7 \\ x_3 \\ x_5 + x_6 + x_7 \\ x_5 \\ x_6 \\ x_7 \end{pmatrix} = x_3 \begin{pmatrix} 1 \\ 1 \\ 1 \\ 0 \\ 0 \\ 0 \\ 0 \end{pmatrix} + x_5 \begin{pmatrix} 1 \\ 0 \\ 0 \\ 1 \\ 1 \\ 0 \\ 0 \end{pmatrix} + x_6 \begin{pmatrix} 0 \\ 1 \\ 0 \\ 1 \\ 0 \\ 1 \\ 0 \end{pmatrix} + x_7 \begin{pmatrix} 1 \\ 1 \\ 0 \\ 1 \\ 0 \\ 0 \\ 1 \end{pmatrix},$$

i.e., as a linear combination of the four specific solutions written above. A little work will show that every solution can be written in a unique way as a linear combination of these vectors. For example, check that $(0, 1, 1, 1, 1, 0, 0)$ is a solution to the system $HX = \mathbf{0}$. Writing this vector as a linear combination of the four given vectors, we must have $x_3 = x_5 = 1$ and $x_6 = x_7 = 0$, so

$$\begin{pmatrix} 0 \\ 1 \\ 1 \\ 1 \\ 1 \\ 0 \\ 0 \end{pmatrix} = \begin{pmatrix} 1 \\ 1 \\ 1 \\ 0 \\ 0 \\ 0 \\ 0 \end{pmatrix} + \begin{pmatrix} 1 \\ 0 \\ 0 \\ 1 \\ 1 \\ 0 \\ 0 \end{pmatrix}$$

is a sum of two of the four given vectors, and can be written in no other way in terms of the four basis vectors.

This example indicates the general fact that for a homogeneous system $AX = \mathbf{0}$, the number of variables not corresponding to leading 1's (i.e., those above that could take on arbitrary values in $\mathbb{Z}_2$) is equal to the dimension of the nullspace of A. Let us call these variables *free variables* and the other variables (of which there are exactly the rank of A) *basic variables*. From the row reduced form of A, the basic variables can be expressed in terms of the free variables. Mimicking the example above, one obtains a distinguished set of solutions to $AX = 0$ by successively setting one free variable equal to 1 and the rest equal to 0. Then any solution can be written uniquely as a linear combination of these solutions. In particular this distinguished set of solutions is a basis for the nullspace of A and therefore, the number of free variables is equal to the dimension of the nullspace. Since every variable is either basic or free and the total number of variables is the number of columns of the matrix, we have the important rank-nullity theorem. The *nullity* of a matrix A is the dimension of the nullspace of A.

Theorem 2.19. *Let A be an $m \times n$ matrix. Then n is equal to the sum of the rank of A and the nullity of A.*

The point of this theorem is that once you know the rank of A, the nullity of A can be immediately calculated. Since we are working over $\mathbb{Z}_2$, the number of solutions to $AX = \mathbf{0}$ is then $2^{\text{nullity}(A)}$. In coding theory this will allow us to determine the number of codewords in a given code.

2.3 The Hamming Code

The Hamming code, discovered independently by Hamming and Golay, was the first example of an error correcting code. Let

$$H = \begin{pmatrix} 0\,0\,0\,1\,1\,1\,1 \\ 0\,1\,1\,0\,0\,1\,1 \\ 1\,0\,1\,0\,1\,0\,1 \end{pmatrix}$$

be the Hamming matrix, described in Example 2.17 above. Note that the columns of this matrix give the base 2 representation of the integers 1–7. The Hamming code C of length 7 is the nullspace of H. More precisely,

$$C = \left\{ v \in \mathbb{Z}_2^7 : Hv^T = \mathbf{0} \right\}.$$

(The transpose is used here because codewords are typically written horizontally, i.e., as row vectors, but without commas to separate the entries). Just as the redundant check digit in an identification number enables the detection of certain errors by the failure of a certain dot product to result in 0, we will see that a code defined as the nullspaces of a matrix can introduce enough redundancies to enable the correction of certain errors.

Before proceeding to this topic, we use Gaussian elimination to gain more detailed information about the Hamming code. Solving as above the linear system $Hx = \mathbf{0}$, we obtain the solution

$$\begin{pmatrix} x_1 \\ x_2 \\ x_3 \\ x_4 \\ x_5 \\ x_6 \\ x_7 \end{pmatrix} = x_3 \begin{pmatrix} 1 \\ 1 \\ 1 \\ 0 \\ 0 \\ 0 \\ 0 \end{pmatrix} + x_5 \begin{pmatrix} 1 \\ 0 \\ 0 \\ 1 \\ 1 \\ 0 \\ 0 \end{pmatrix} + x_6 \begin{pmatrix} 0 \\ 1 \\ 0 \\ 1 \\ 0 \\ 1 \\ 0 \end{pmatrix} + x_7 \begin{pmatrix} 1 \\ 1 \\ 0 \\ 1 \\ 0 \\ 0 \\ 1 \end{pmatrix}.$$

Therefore, C has dimension 4, and the set $\{1110000, 1001100, 0101010, 1101001\}$ forms a basis for C (we will discuss these terms more rigorously in Chap. 4). If one were to write out all 16 codewords in C, one would find the distance of C to be exactly 3.

Linear codes like C are identified by their length, dimension, and minimum distance. Thus C is referred to as a $(7, 4, 3)$-code, because its length is 7, its dimension is 4, and its minimum distance is equal to 3. In particular, we deduce from Proposition 2.8 that C corrects 1 error.

The code C has a particularly elegant decoding algorithm, which we now describe. Let $\{e_1, \ldots, e_7\}$ be the standard basis for $\mathbb{Z}_2^7$. We point out a fact of matrix multiplication: He_i^T is equal to the ith column of H. Moreover, we note that the seven nonzero vectors in $\mathbb{Z}_2^3$ are exactly the seven columns of H.

Suppose that v is a codeword that is transmitted as a word $w \neq v$ and that exactly one error has been made in transmission. Then $w = v + e_i$ for some i. However, we do not yet know i, so we cannot yet determine v from w. However,

$$Hw^T = H(v + e_i)^T = Hv^T + He_i^T = He_i^T,$$

and He_i^T is the ith column of H, as we pointed out above. Therefore i is determined by computing Hw^T and comparing the result with the columns of H. The column number of H given by Hw^T is exactly i. Then w is decoded to $w + e_i$, which must be equal to v since we assumed that only one error was made in transmission. To summarize this error correcting algorithm: Given a word w, calculate Hw^T. If the product is 0, then w is a codeword. If it is not, then it is equal to the ith column of H for a unique integer i. Then $w + e_i$ is a valid codeword, and is the closest codeword to w.

The Hamming code C has an additional property: every word is within distance 1 of a codeword. To see this, suppose that w is a word. If $Hw^T = \mathbf{0}$, then w is a codeword. If not, then Hw^T is a nonzero 3-tuple. Therefore, it is equal to a column of H; say that Hw^T is equal to the ith column of H. Then $Hw^T = He_i^T$, so $H(w^T + e_i^T) = \mathbf{0}$, so that $w + e_i \in C$. The word $w + e_i$ is then a codeword a distance of 1 from w. A code that corrects t errors and for which every word is within t of some codeword is called *perfect*. Such codes are particularly nice, in part because a decoding procedure will always return a codeword. Later we will see some important codes that are not perfect. So perfection is not the ultimate goal. Nevertheless, we can be inspired by the words of Lord Chesterfield: "Aim at perfection in everything, though in most things it is unattainable. However, they who aim at it, and persevere, will come much nearer to it than those whose laziness and despondency make them give it up as unattainable."

Exercises

1. Let C be the code (of length n) of solutions to a matrix equation $Ax = \mathbf{0}$. Define a relation on the set $\mathbb{Z}_2^n$ of words of length n by $u \equiv v \bmod C$ if $u + v \in C$. Prove that this is an equivalence relation, and that for any word w, the equivalence class of w is the coset $C + w$.
2. Verify that 1100110 belongs to the $(7, 4, 3)$ Hamming code.
3. 1011110 is not a codeword for the $(7, 4, 3)$ Hamming code. Use the decoding algorithm above to identify the error and to correct it.
4. Consider the matrix $\hat{H}$ with entries in $\mathbb{Z}_2$ whose columns consist of the base 2 representations of the integers from 1 though 15 in increasing order. Determine the rank of $\hat{H}$ and find a basis for its nullspace.
5. Find the minimum distance and error correction capability of the nullspace of $\hat{H}$ defined in the previous problem. Is this code perfect?

2.4 Coset Decoding

To apply MLD (Maximum Likelihood Decoding, Sect. 2.1) what we must do, given a received word w, is search through all the codewords to find the codeword c closest to w. This can be a slow and tedious process. There are more efficient methods, assuming the code is built in a manner similar to that of the Hamming code, i.e., that the code C is given as the nullspace of an $m \times n$ matrix H:

$$C = \left\{v \in \mathbb{Z}_2^n : Hv^T = \mathbf{0}\right\}$$

and therefore has length n and dimension equal to the nullity of H. We fix the symbols C and H to have this meaning in this section.

Definition 2.20. Let w be a word. Then the coset $C + w$ of w is the set $\{c + w : c \in C\}$.

Recall two facts about C. First, by the definition of C, the zero vector $\mathbf{0}$ is an element of the code, since $H\mathbf{0} = \mathbf{0}$. From this we see that $w \in C + w$, since $w = \mathbf{0} + w$. Second, if $u, v \in C$, our assumption of linearity requires that $u + v \in C$ (i.e., $H(u + v)^T = Hu^T + Hv^T = \mathbf{0} + \mathbf{0} = \mathbf{0}$).

We now discuss an important property of cosets, namely that any two cosets are either equal or are disjoint. In fact cosets are the equivalence classes for the following equivalence relation defined on $\mathbb{Z}_2^n$:

Two words x and y are related if $x + y \in C$.

We write $x \sim y$ when this occurs. To see that this is an equivalence relation, we must verify the three properties of reflexivity, symmetry, and transitivity. For reflexivity, recall that addition in $\mathbb{Z}_2^n$ is componentwise so for every x in $\mathbb{Z}_2^n$ we have $x + x = \mathbf{0}$, which is an element of C. Thus $x \sim x$. Next, suppose that $x \sim y$. To verify symmetry, we must show that $y \sim x$. The assumption that $x \sim y$ means $x + y \in C$. However, $x + y = y + x$; therefore, since $y + x \in C$, we have $y \sim x$. Finally, for transitivity, suppose that $x \sim y$ and $y \sim z$. Then $x + y \in C$ and $y + z \in C$. Adding these codewords results in a codeword by the previous paragraph. However,

$$(x + y) + (y + z) = x + (y + y) + z = x + \mathbf{0} + z = x + z,$$

by the properties of vector addition. Since the result, $x + z$, is an element of C, we have $x \sim z$, as desired. So ~ is an equivalence relation.

The equivalence class of a word x is

$$\begin{aligned}\{y : y \sim x\} &= \{y : x + y \in C\} = \{y : y = c + x \text{ for some } c \in C\} \\ &= C + x.\end{aligned}$$

The third equality follows since if $x + y = c$, then $y = c + x$.

Proposition 2.21. *If x and y are words, then $C + x = C + y$ if and only if $Hx^T = Hy^T$.*

Proof. Suppose first that $C + x = C + y$. Then $x \sim y$, so $x + y \in C$. By definition of C, we have $H(x + y)^T = \mathbf{0}$. Expanding the left-hand side, and using the fact that $(x + y)^T = x^T + y^T$, we get $Hx^T + Hy^T = \mathbf{0}$, so $Hx^T = Hy^T$. Conversely, suppose that $Hx^T = Hy^T$. Then

$Hx^T + Hy^T = \mathbf{0}$, or $H(x+y)^T = \mathbf{0}$. This last equation says $x + y \in C$, and so $x \sim y$. From this relation between x and y, we obtain $C + x = C + y$, since these are the equivalence classes of x and y, and these classes are equal since x and y are related. □

Example 2.22. Let

$$H = \begin{pmatrix} 1\,1\,1\,1 \\ 1\,1\,0\,0 \end{pmatrix}.$$

A short calculation shows that $C = \{0000, 1100, 0011, 1111\}$. The cosets of C are then seen to be

$$\begin{aligned} C + 0000 &= \{0000, 1100, 0011, 1111\}, \\ C + 1000 &= \{1000, 0100, 1011, 0111\}, \\ C + 0010 &= \{0010, 1110, 0001, 1101\}, \\ C + 1010 &= \{1010, 0110, 1001, 0101\}. \end{aligned}$$

We also point out that $C = C + 0000 = C + 1100 = C + 0011 = C + 1111$; in other words, $C = C + v$ for any $v \in C$. Each coset in this example is equal to the coset of four vectors, namely the four vectors in the coset.

Introducing some coding theory terminology, call Hx^T the *syndrome* of x. Syndromes enable more efficient decoding. Suppose that a word w is received. If c is the closest codeword to w, let $e = c + w$. Then e is the *error word*, in that e has a digit equal to 1 exactly when that digit was transmitted incorrectly in c. Note that e is the word of smallest possible weight of the form $v + w$ with $v \in C$ since $\mathrm{wt}(e) = D(c, w)$. If we can determine e, then we can determine c by $c = e + w$. To see where the syndrome comes into play, multiply both sides of the equation $e^T = c^T + w^T$ by H to obtain

$$\begin{aligned} H^eT &= H(c+w)^T = Hc^T + Hw^T = \mathbf{0} + Hw^T \\ &= Hw^T \end{aligned}$$

which is the syndrome of the received word. We therefore compute He^T by computing Hw^T. Proposition 2.21 says that $C + e = C + w$; in other words, $e \in C + w$. More generally, any pair of words with the same syndrome determine the same coset of C. Since c is the closest codeword to w, the word e is then the word of least weight in the coset $C + w$. We then find e by searching the words in $C + w$ for the word of least weight; such a word is called a *coset leader*. To decode with cosets, we compute and list a coset leader for each coset (i.e., syndrome).

Example 2.23. Let

$$H = \begin{pmatrix} 1\,1\,0\,0\,0 \\ 1\,0\,1\,1\,0 \\ 1\,0\,1\,0\,1 \end{pmatrix}.$$

Then $C = \{00000, 11100, 00111, 11011\}$. We see that the distance of C is 3, so C is 1-error correcting. The cosets of C are

$$\begin{gathered}
\{\mathbf{00000}, 00111, 11011, 11100\}, \\
\{01110, \mathbf{10010}, 01001, 10101\}, \\
\{\mathbf{00010}, 00101, 11001, 11110\}, \\
\{11111, 11000, 00011, \mathbf{00100}\}, \\
\{01111, \mathbf{01000}, 10100, 10011\}, \\
\{01101, 10110, \mathbf{01010}, 10001\}, \\
\{01100, \mathbf{10000}, 10111, 01011\}, \\
\{11010, \mathbf{00001}, 11101, 00110\}.
\end{gathered}$$

By searching through each of the eight cosets (a word of minimal weight in each coset has been boldfaced), we can then build the following syndrome table:

Syndrome	**Coset leader**
000	00000
101	10010
010	00010
011	00100
100	01000
110	01010
111	10000
001	00001

The following examples illustrate the use of a syndrome table for decoding. Suppose that $w = 10010$ is received. Calculating $(Hw^T)^T = wH^T$ results in 101. First of all, since $Hw^T \neq \mathbf{0}$, and by the definition of the code as the nullspace of H, the vector w is not a codeword. From the syndrome table, we see that 101 is the second syndrome listed. The corresponding coset leader is $e = 10010$. The received word w is decoded as $c = w + e = 00000$. Similarly, if we receive the word $w = 11111$, we calculate $wH^T = 011$. The corresponding coset leader is $e = 00100$, so the corrected codeword is $e + w = 11011$.

Clearly using the syndrome table requires much less computation than checking the distance between w and all 16 codewords to find the closest one. The fact that choices of the weight 2 coset leader were made for syndromes 110 and 101 shows that this code cannot correct two errors and also that it is not perfect.

Exercises

1. Let C be the code consisting of all solutions of the matrix equation $Ax^T = \mathbf{0}$, where

$$A = \begin{pmatrix} 0\,0\,0\,1\,1\,1\,1\,0 \\ 0\,1\,1\,0\,0\,1\,1\,0 \\ 1\,0\,1\,0\,1\,0\,1\,0 \\ 1\,1\,1\,1\,1\,1\,1\,1 \end{pmatrix}.$$

a. Calculate C and determine its distance and error correcting capability.
b. Construct the syndrome table for C.
c. Use the table to decode the vectors 10101101, 01011011, and 11000000.

2. List all of the cosets of the code $C = \{00000, 11100, 00111, 11011\}$.
3. Find the cosets of the Hamming code.
4. Let C be the code consisting of solutions to $Ax^T = \mathbf{0}$, where

$$A = \begin{pmatrix} 1\,1\,1\,0\,0\,0 \\ 0\,1\,1\,1\,0\,0 \\ 0\,0\,0\,1\,1\,1 \\ 1\,0\,0\,0\,1\,1 \end{pmatrix}.$$

Build the syndrome table for C. Determine the distance of C. Use it to decode, if possible, 111110 and 100000. Feel free to use the Maple worksheet Cosets.mw.

2.5 The Golay Code

In this section we discuss a length 24 code used by NASA in the 1970s and 1980s to transmit images of Jupiter and Saturn photographed by the Voyager spacecraft. This code, called the *extended Golay code,* is the set of solutions to the matrix equation $Hx^T = \mathbf{0}$, where H is the 12×24 matrix $H = [I \mid B]$ whose left half is the 12×12 identity matrix I and whose right half is the symmetric 12×12 matrix

$$B = \begin{pmatrix} 1\,1\,0\,1\,1\,1\,0\,0\,0\,1\,0\,1 \\ 1\,0\,1\,1\,1\,0\,0\,0\,1\,0\,1\,1 \\ 0\,1\,1\,1\,0\,0\,0\,1\,0\,1\,1\,1 \\ 1\,1\,1\,0\,0\,0\,1\,0\,1\,1\,0\,1 \\ 1\,1\,0\,0\,0\,1\,0\,1\,1\,0\,1\,1 \\ 1\,0\,0\,0\,1\,0\,1\,1\,0\,1\,1\,1 \\ 0\,0\,0\,1\,0\,1\,1\,0\,1\,1\,1\,1 \\ 0\,0\,1\,0\,1\,1\,0\,1\,1\,1\,0\,1 \\ 0\,1\,0\,1\,1\,0\,1\,1\,1\,0\,0\,1 \\ 1\,0\,1\,1\,0\,1\,1\,1\,0\,0\,0\,1 \\ 0\,1\,1\,0\,1\,1\,1\,0\,0\,0\,1\,1 \\ 1\,1\,1\,1\,1\,1\,1\,1\,1\,1\,1\,0 \end{pmatrix}$$

which satisfies $B^2 = I$.

The photographs were made using 4,096 colors. Each color was encoded with a codeword from the Golay code. By solving the matrix equation $Hx^T = \mathbf{0}$, we can see that there are indeed 4,096 codewords. Furthermore, a tedious check of all codewords shows that the distance of the Golay code has distance $d = 8$. Thus, the code can correct $\lfloor (8-1)/3 \rfloor = 3$ errors, hence up to three out of the 24 digits of a codeword can be corrupted and still the original information will be retrievable.

Because this code can correct more than one error, any decoding procedure is bound to be more complicated than that for the Hamming code. We give a decoding procedure based on some simple facts about the matrix B. Its validity is left to a series of homework problems.

To make it more convenient to work with this code, we write a word $u = (u_1, u_2)$, where u_1 consists of the first 12 digits and u_2 the remaining 12. Since $H = [I \mid B]$, we see that $u \in C$ if and only if

$Hu^T = \mathbf{0}$, which is true if and only if $u_1^T + Bu_2^T = \mathbf{0}$. For a received word w, the following steps are performed to decode w. We write v for the codeword to be determined from w. As usual, e_i denotes the 12-tuple with ith-entry 1 and all other entries 0, while b_i denotes the ith row of the matrix B.

1. Compute $s^T = Hw^T$. If $s^T = \mathbf{0}$, then w is a codeword.
2. If $1 \leq \text{wt}(s) \leq 3$, then $v = w + (s, \mathbf{0})$.
3. If $\text{wt}(s) > 3$ and $\text{wt}(s + b_i) \leq 2$ for some i, then $v = w + (s + b_i, e_i)$.
4. If we haven't yet determined v, then compute sB, which is equal to $(Bs^T)^T$ by symmetry of B.
5. If $1 \leq \text{wt}(sB) \leq 3$, then $v = w + (0, sB)$.
6. If $\text{wt}(sB) > 3$ and $wt(sB + b_i) \leq 2$ for some i, then $v = w + (e_i, sB + b_i)$.
7. If we haven't determined v, then w cannot be decoded.

Example 2.24. Suppose that $w = 001001001101101000101000$ is received. We calculate $s^T = Hw^T$, and find $s = 110001001001$ with $\text{wt}(s) = 5$. We see that $\text{wt}(s + b_5) = 2$. Therefore, by Step 3, w is decoded as $v = w + (s + b_5, e_5) = w + (000000010010, 000010000000) = 001001011111101010101000$.

Exercises

For these problems, some of the theoretical facts behind the decoding procedure for the Golay code are verified. We use the following setup: C is the Golay code, H is the 12×24 matrix $[I \mid B]$ mentioned in the text, w is a received word, $s^T = Hw^T$. Our conventions are that a 24-tuple written as (u_1, u_2) means that each u_i is a 12-tuple and that the ith row (and column) of the symmetric matrix B is denoted by b_i. Let v be the closest codeword to w and write $v = w + e$. Since the Golay code is asserted to be 3-error correcting, we assume that $\text{wt}(e) \leq 3$.

Recall that $B^2 = I$ and $B^T = B$. A straightforward but tedious check of the rows of B shows that (i) $\text{wt}(b_i) \geq 7$ for all i; (ii) $\text{wt}(b_i + b_j) \geq 6$ if $i \neq j$; (iii) $\text{wt}(b_i + b_j + b_k) \geq 5$ for all i, j, k. Since $B^T = B$, the ith column of B is b_i, and so $Be_i = b_i$. You are free to use these facts.

1. Suppose that $e = (u, \mathbf{0})$; with $\text{wt}(u) \leq 3$. Show that $s = u$, and conclude that $v = w + (s, \mathbf{0})$.
2. Suppose that $e = (u, e_i)$ with $\text{wt}(u) \leq 2$. Show that $s = u + b_i$. Conclude that $\text{wt}(s) > 3$ and $\text{wt}(s + b_i) \leq 2$, and that $v = w + (s + b_i, e_i)$.
3. Suppose that $e = (\mathbf{0}, u)$ with $\text{wt}(u) \leq 3$. Show that s is the sum of at most three of the b_i and that $u = sB$. Conclude that $\text{wt}(s) > 3$ but $\text{wt}(Bs) \leq 3$, and that $v = w + (\mathbf{0}, sB)$.
4. Suppose that $e = (e_i, u)$ with $\text{wt}(u) \leq 2$. Show that $s = e_i + uB$, and that $sB = b_i + u$. Conclude that $\text{wt}(s) > 3$, $\text{wt}(s + b_i) > 2$ for any i, and that $e = (e_i, sB + b_i)$, so $v = w + (e_i, sB + b_i)$.

These four problems show, given any possibility of an error vector e having weight at most 3, how we can determine it in terms of the syndrome s. Reading these four problems backwards yields the decoding procedure discussed in this section.

References

1. Hankerson DC et al (2000) Coding theory and cryptography: the essentials, 2nd edn. Marcel Dekker, New York
2. Herstein I (1975) Topics in algebra, 2nd edn. Wiley, Hoboken
3. Talbot J, Welsh D (2006) Complexity and cryptography: an introduction. Cambridge University Press, Cambridge

Chapter 3
Rings and Fields

We are all familiar with the natural, rational, real, and complex number systems and their arithmetic, but other mathematical systems exhibit similar arithmetic properties. The previous chapter, for instance, introduced the set of integers modulo n, and its addition, subtraction, and multiplication. In high school algebra you worked with polynomials, and saw how to add, subtract, and multiply them. In linear algebra you saw how arithmetic operations are performed on matrices, and might have seen vector spaces, with their addition, subtraction, and scalar multiplication. Many of the functions you studied in precalculus and calculus can be combined by addition, subtraction, multiplication, division, and also composition.

There are many similarities among these different systems. The idea of abstract algebra is to distill the algebraic properties common to systems arising naturally in mathematics, and then investigate general systems having these properties, without specific reference to any particular one. We will study several different algebraic systems, the first of which, that of a ring, generalizes the integers and the integers modulo n. Looking back at the list of properties that the modular operations satisfy, notice that analogues of these properties hold in the familiar number systems listed above. These properties, except for commutativity of multiplication, hold also for matrix operations and occur in so many contexts that it is necessary to study systems that satisfy them.

3.1 The Definition of a Ring

To begin, we must first formalize the meaning of an operation. With integer addition, one starts with two integers and the process of adding them returns a single integer. Therefore, addition is a function that takes as input a pair of integers and gives as output a single integer. Multiplication can be viewed similarly. These are special examples of binary operations. Recall that the Cartesian product $A \times B$ of two sets A and B is the set of all pairs (a, b) with $a \in A$ and $b \in B$. In other words,

$$A \times B = \{(a,b) : a \in A, b \in B\}.$$

Definition 3.1. If S is a set, then a binary operation on S is a function from $S \times S$ to S.

Example 3.2. The operations of addition, subtraction, and multiplication on the various number systems are all examples of binary operations. Division is not an example. Division

D.R. Finston and P.J. Morandi, *Abstract Algebra: Structure and Application*,
Springer Undergraduate Texts in Mathematics and Technology, DOI 10.1007/978-3-319-04498-9_3

on the real numbers is not a function from $\mathbb{R} \times \mathbb{R}$ to $\mathbb{R}$ but a function from $\mathbb{R} \times (\mathbb{R} - \{0\})$ to $\mathbb{R}$. In other words, since we cannot divide by 0, division is not defined on pairs of the form $(a, 0)$.

Example 3.3. If T is any set, then union $\cup$ and intersection $\cap$ are examples of binary operations on the set of all subsets of T. That is, given any two subsets of T, the union and intersection of the sets is again a subset of T.

The properties of a specific binary operation $f : S \times S \to S$ on a particular set S can be expressed in terms of the function f. For instance, if f was a commutative binary operation, then $f(a, b) = f(b, a)$ for every pair of elements $(a, b) \in S \times S$. Associativity would be expressed as $f(a, f(b, c)) = f(f(a, b), c)$ for every triple of elements $(a, b, c) \in S \times S \times S$. This notation can quickly become cumbersome so we will rarely use it, especially since the operations that arise most frequently in practice have more natural and familiar expressions.

Example 3.4. Here is an example from multivariable calculus. Consider the set $\mathbb{R}^3$ of 3-tuples of real numbers. The cross product is a binary operation on $\mathbb{R}^3$ given for 3-tuples (a_1, a_2, a_3) and (b_1, b_2, b_3) by the formula

$$\begin{aligned}(a_1, a_2, a_3) \times (b_1, b_2, b_3) &= \det \begin{pmatrix} i & j & k \\ a_1 & a_2 & a_3 \\ b_1 & b_2 & b_3 \end{pmatrix} \\ &= (a_2 b_3 - a_3 b_2) i + (a_3 b_1 - a_1 b_3) j + (a_1 b_2 + a_2 b_1) k,\end{aligned}$$

where i, j, and k are alternative notations for the usual unit vectors e_1, e_2, and e_3. The reason for giving this example here is that it provides a natural example of a non-associative operation. For instance,

$$(e_1 \times e_1) \times e_2 = 0 \times e_2 = 0$$

while

$$e_1 \times (e_1 \times e_2) = e_1 \times e_3 = -e_2.$$

Thus, the associative property can't be expected to hold for every naturally occurring binary operation.

While it is possible to investigate sets with operations in complete generality, most useful algebraic structures do satisfy at least some of our familiar properties, such as the commutative and associative laws. In particular, with the notion of a binary operation on a set, we can give the definition of a ring. This is the structure that underlies most of the examples mentioned above and will be of importance in what follows.

Definition 3.5. A ring is a nonempty set R together with two binary operations $+$ and $\cdot$ such that, for all $a, b, c \in R$, the following properties hold:

- For each $a, b \in R$, both $a + b$ and $a \cdot b$ are elements of R.
- Commutativity of addition: $a + b = b + a$.
- Associativity of addition: $a + (b + c) = (a + b) + c$.
- Existence of an additive identity: There is an element $0 \in R$ with $a + 0 = a$ for every $a \in R$.
- Existence of additive inverses: For each $a \in R$ there is an element $s \in R$ with $a + s = 0$.
- Associativity of multiplication: $a \cdot (b \cdot c) = (a \cdot b) \cdot c$.
- Left distributivity: $a \cdot (b + c) = a \cdot b + a \cdot c$.

- Right distributivity: $(b + c) \cdot a = b \cdot a + c \cdot a$.
- Existence of a multiplicative identity: There is an element $1 \in R$ with $1 \cdot a = a \cdot 1 = a$ for all $a \in R$.

We tacitly assume that $0 \neq 1$ in a ring R. Otherwise R can only be the trivial ring $\{0\}$ (see Exercise 14 below). The example above of $\mathbb{R}^3$ with vector addition as the addition and the cross product as multiplication does not satisfy our definition of a ring because of the failure of associativity of the cross product and the absence of an identity for this operation. Algebraic structures with nonassociative multiplication do have importance throughout mathematics, but these lie beyond the scope of this book.

Before we give examples of rings, we point out a few things about the definition. The first property is simply a restatement of what it means for $+$ and $\cdot$ to be binary operations. One often refers to these properties as closure under addition and under multiplication. Next, we did not include commutativity of multiplication in this list. If a ring R satisfies $a \cdot b = b \cdot a$ for all $a, b \in R$, then we call R a *commutative ring*. If a ring is commutative, then the two distributivity laws reduce to the same thing, and the requirement that $1 \cdot a = a \cdot 1$ is redundant. The definition of a ring does not include the existence of multiplicative inverses of all nonzero elements. Indeed, for a composite integer n, we have seen in $\mathbb{Z}_n$ that elements $\overline{a} \neq \overline{0}$ need not have multiplicative inverses. The existence of multiplicative inverses is an important issue that will be addressed in several examples.

Example 3.6. The set $\mathbb{Z}$ of integers forms a ring under the usual addition and multiplication operations. The defining properties of a ring are well known to hold for $\mathbb{Z}$. The set $\mathbb{Q}$ of rational numbers also forms a ring under the usual operations. So do the set $\mathbb{R}$ of real numbers and the set $\mathbb{C}$ of complex numbers. All four of these are, of course, commutative rings.

Example 3.7. The set $M_n(\mathbb{R})$ of $n \times n$ matrices with real number entries under matrix addition and multiplication forms a ring. It is proved in linear algebra courses that the operations of matrix addition and multiplication satisfy the properties above. For instance, the zero matrix in $M_2(\mathbb{R})$

$$\begin{pmatrix} 0 & 0 \\ 0 & 0 \end{pmatrix}$$

is the additive identity, and the identity matrix

$$\begin{pmatrix} 1 & 0 \\ 0 & 1 \end{pmatrix}$$

is the multiplicative identity. This ring is not commutative. For example,

$$\begin{pmatrix} 1 & 0 \\ 1 & 1 \end{pmatrix} \begin{pmatrix} 0 & 1 \\ 2 & 1 \end{pmatrix} = \begin{pmatrix} 0 & 1 \\ 2 & 2 \end{pmatrix}$$

while

$$\begin{pmatrix} 0 & 1 \\ 2 & 1 \end{pmatrix} \begin{pmatrix} 1 & 0 \\ 1 & 1 \end{pmatrix} = \begin{pmatrix} 1 & 1 \\ 3 & 1 \end{pmatrix},$$

so the order of multiplication matters.

Example 3.8. If n is a positive integer, then the set $\mathbb{Z}_n$ of integers modulo n, with addition and multiplication of residues classes defined as in Chapter I, is a ring. In fact the ring properties for $\mathbb{Z}_n$ were verified there. Since the multiplication is commutative, $\mathbb{Z}_n$ is furthermore a commutative ring.

Example 3.9. Let R be the set of all continuous (real-valued) functions defined on the interval $[0, 1]$ and consider the binary operations of function addition and multiplication. Recall that these function operations are defined pointwise. That is, if f and g are functions, then $f + g$, $f - g$, and fg are defined by

$$\begin{aligned}(f + g)(x) &= f(x) + g(x),\\ (f - g)(x) &= f(x) - g(x),\\ (fg)(x) &= f(x)g(x).\end{aligned}$$

In calculus one shows that the sum, difference, and product of continuous functions are again continuous. Thus, we have operations of addition and multiplication on the set R. It is possible that the ring properties were verified in your calculus class. We do not verify all here, but limit ourselves to discussing some. The additive identity of R is the zero function 0 which is defined by $0(x) = 0$ for all $x \in [0, 1]$. The multiplicative identity of R is the constant function 1 defined by $1(x) = 1$ for all $x \in [0, 1]$. Commutativity of addition holds because if $f, g \in R$ and $x \in [0, 1]$, then

$$\begin{aligned}(f + g)(x) &= f(x) + g(x)\\ &= g(x) + f(x)\\ &= (g + f)(x).\end{aligned}$$

Thus, the functions $f + g$ and $g + f$ agree at every value in the domain, which means, by definition, that $f + g = g + f$. To prove this we used the definition of function addition and commutativity of addition of real numbers. All of the other ring properties follow from the definition of the operations and appropriate ring properties of $\mathbb{R}$. Furthermore, multiplication of functions is commutative, so the ring of continuous functions on $[0, 1]$ is commutative.

Example 3.10. Let $\mathbb{R}[x]$ be the set of all polynomials with real number coefficients. While the operations of addition and multiplication of polynomials are no doubt familiar, we recall their definitions now, primarily to give a simple notation for them which will be useful in later developments. First note that given two polynomials $f(x) = \sum_{i=1}^{n} a_i x^i$ and $g(x) = \sum_{i=1}^{m} b_i x^i$ we can add on terms with zero as coefficients in order to be able to assume that $m = n$. With this in mind, we have

$$(f + g)(x) = \sum_{i=1}^{n} a_i x^i + \sum_{i=1}^{n} b_i x^i = \sum_{i=1}^{n} (a_i + b_i) x^i .$$

Similarly, by collecting coefficients of like powers of x, we have

$$(f \cdot g)(x) = \sum_{i=0}^{n} a_i x^i \cdot \sum_{i=0}^{m} b_i x_i = \sum_{j}^{n+m} \left(\sum_{i=0}^{j} a_i b_{j-i} \right) x_j .$$

It is likely that in some earlier course you saw that all of the ring properties hold for $\mathbb{R}[x]$, specifically that multiplication of polynomials is commutative and associative, so $\mathbb{R}[x]$ is a commutative ring.

Example 3.11. This example is important in the study of set theory and logic. Let T be a set, and let $\mathcal{R}$ be the set of all subsets of T. We have two binary operations on $\mathcal{R}$, namely union and intersection. However, $\mathcal{R}$ does not form a ring under these two operations. The identity element for union is the empty set ϕ since $A \cup \phi = \phi \cup A = A$ for any $A \in T$. Similarly T serves as the identity element for intersection. On the other hand one can check that neither operation allows for inverses: given a nonempty subset A of T, what subset B can satisfy $A \cup B = \phi$? And if A is a proper subset of T what subset B can satisfy $A \cap B = T$? In particular, neither union nor intersection can be considered as addition. However, we can introduce new operations to come up with a ring. Define addition and multiplication on $\mathcal{R}$, respectively, by

$$A + B = (A - B) \cup (B - A) = (A \cup B) - (A \cap B),$$
$$A \cdot B = A \cap B.$$

Then $\mathcal{R}$, together with these operations, does form a commutative ring. We leave the details of the proof to an exercise. We do point out some interesting properties of this ring. First of all, $0 = \varnothing$ for this ring since $A + \varnothing = (A \cup \varnothing) - (A \cap \varnothing) = A - \varnothing = A$ for every subset A of T. Also, $1 = T$ since $A \cdot T = A \cap T = A$ for every subset A. Next, for any A we have $A + A = (A \cup A) - (A \cap A) = A - A = \varnothing$. Therefore, $-A = A$ for any A! This would seem to be a very unusual property, although we will see it frequently in the sections on coding theory. Finally, note that $A \cdot A = A \cap A = A$.

Example 3.12. In this example we describe a method of constructing a new ring from two existing rings. Let R and S be rings. Define operations $+$ and $\cdot$ on the Cartesian product $R \times S$ by

$$(r, s) + (r', s') = (r + r', s + s'),$$
$$(r, s) \cdot (r', s') = (rr', ss')$$

for all $r, r' \in R$ and $s, s' \in S$, where $r + r'$ and $s + s'$ are calculated in the rings R and S, respectively, and similarly for rr' and ss'. Then one can verify that $R \times S$, together with these operations, is indeed a ring. If 0_R and 0_S are the additive identities of R and S, respectively, then $(0_R, 0_S)$ is the additive identity for $R \times S$. Similarly, if 1_R and 1_S are the multiplicative identities of R and S, respectively, then $(1_R, 1_S)$ is the multiplicative identity of $R \times S$.

3.2 First Properties of Rings

There are some properties with which you are very familiar for the ordinary number systems that hold for any ring. First of all, the additive and multiplicative identities are unique. We leave one of these facts for homework, and prove the other one now.

Lemma 3.13. *Let R be a ring. The additive identity is unique.*

Proof. Suppose there are elements 0 and $0'$ in R that satisfy $a + 0 = a$ and $a + 0' = a$ for all $a \in R$. Recalling that addition is commutative, we have $a + 0 = 0 + a = a$ and $a + 0' = 0' + a = a$ for any a. If we use the first pair of equations for $a = 0'$, we get $0' + 0 = 0'$.

On the other hand, if we use the second pair with $a = 0$, we get $0' + 0 = 0$. Thus, $0 = 0'$. This shows that there can be only one additive identity for R. □

We next remark that the additive inverse of any element is unique (the proof is Exercise 1 below). Since the additive inverse of a ring element is unique, we use the familiar notation $-a$ to represent the additive inverse of a. The additive inverse enables the definition of subtraction in the usual manner: $a - b = a + (-b)$.

Lemma 3.14. *Let R be a ring, and let $a, b \in R$.*

1. $-(-a) = a$;
2. $-(a + b) = -a + (-b)$;
3. $-(a - b) = -a + b$.

Proof. Let $a \in R$. By definition of $-a$, we have $a + (-a) = (-a) + a = 0$. This equation tells us that a is the additive inverse of $-a$. That is, $a = -(-a)$, which proves the first property. For the second property, let $a, b \in R$. Then $-(a + b)$ is the additive inverse of $a + b$. We show $-a + (-b)$ is also the additive inverse of $a + b$. To do this, we calculate

$$\begin{aligned}
(a + b) + (-a + (-b)) &= ((a + b) + (-a)) + (-b) \\
&= (a + (b + (-a)) + (-b) \\
&= (a + ((-a) + b)) + (-b) \\
&= ((a + (-a)) + b) + (-b) \\
&= (0 + b) + (-b) \\
&= b + (-b) \\
&= 0.
\end{aligned}$$

Each of these steps followed from one of the properties in the definition of a ring. Because we have shown that $(a + b) + (-a + (-b)) = 0$, the element $-a + (-b)$ is the additive inverse of $a+b$. This tells us that $-a+(-b) = -(a+b)$. Finally, for the third property, again let $a, b \in R$. We use a similar argument as for part 2, leaving out a few steps which you are encouraged to fill in. We have

$$\begin{aligned}
(a - b) + (-a + b) &= (a + (-b)) + (-a + b) \\
&= (a + (-b) + (-a)) + b \\
&= (a + (-a) + (-b)) + b \\
&= (0 + (-b)) + b \\
&= (-b) + b \\
&= 0
\end{aligned}$$

Therefore, $-a + b = -(a - b)$. □

Proposition 3.15 (Cancellation Law of Addition). *Let $a, b, c \in R$. If $a + b = a + c$, then $b = c$.*

Proof. Let $a, b, c \in R$, and suppose that $a + b = a + c$. By adding $-a$ to both sides, we get the following string of equalities.

$$\begin{aligned} a+b &= a+c \\ -a+(a+b) &= -a+(a+c) \\ (-a+a)+b &= (-a+a)+c \\ 0+b &= 0+c \\ b &= c \end{aligned}$$

Again, notice that in each line we used properties from the definition of a ring. □

The element 0 is defined with respect to addition, but we want to know how it behaves with respect to multiplication. The distributive properties are the only properties in the definition of a ring that involve both operations. They will then be necessary to use to prove anything that relates the two operations. Such an example is the following lemma:

Lemma 3.16. *Let R be a ring. If $a \in R$, then $a \cdot 0 = 0 \cdot a = 0$.*

Proof. Let $a \in R$. To prove that $a \cdot 0 = 0$, we use additive cancellation together with the definition of an additive identity. We have

$$\begin{aligned} a \cdot 0 + a = a \cdot 0 + a \cdot 1 &= a \cdot (0+1) \\ &= a \cdot 1 \\ &= a \\ &= 0 + a. \end{aligned}$$

Subtracting a from both sides yields $a \cdot 0 = 0$. A similar argument shows that $0 \cdot a = 0$. □

Remark 3.17. Lemma 3.16 shows that 0 has no multiplicative inverse in a ring with more than one element (i.e., in a ring in which $0 \neq 1$). Thus "division by 0" is meaningless.

Let us now address the existence of multiplicative inverses. To illustrate an underlying issue, consider an important related concept from high school mathematics that often gets too little attention. First an example: Suppose you wish to solve the equation $x^2 - 5x - 6 = 0$ in the real numbers by factoring. Since $x^2 - 5x - 6 = (x+1)(x-6)$, our equation is equivalent to $(x+1)(x-6) = 0$. Since the real numbers have the property that the product of two nonzero numbers is again nonzero, our problem reduces to $x + 1 = 0$ or $x - 6 = 0$, and hence the two solutions $x = -1$ and $x = 6$.

Factoring is a useful technique in solving equations precisely because of the property of the real numbers that we just used. On the other hand, we have seen for some n, that the ring $\mathbb{Z}_n$ does not possess the property that the product of nonzero elements is nonzero (e.g. in $\mathbb{Z}_6$ we have $\overline{2} \cdot \overline{3} = \overline{0}$). Likewise, this behavior occurs in rings of matrices. For example,

$$\begin{pmatrix} 1 & 0 \\ 0 & 0 \end{pmatrix} \begin{pmatrix} 0 & 0 \\ 0 & 1 \end{pmatrix} = \begin{pmatrix} 0 & 0 \\ 0 & 0 \end{pmatrix}.$$

If we are trying to solve equations in a ring, then it may be crucial to know whether or not the product of nonzero elements can equal to zero.

Definition 3.18. Let R be a ring and a a nonzero element of R. If there is a nonzero element $b \in R$ with $a \cdot b = 0$ then a is said to be a zero divisor.

A useful alternative phrasing of the definition is the contrapositive. A nonzero element $a \in R$ is not a zero divisor if whenever $b \in R$ with $a \cdot b = 0$, then $b = 0$. Terminology about zero divisors varies from textbook to textbook. Some books only define zero divisors for commutative rings. Some books consider 0 to be a zero divisor and others do not. Others talk about left and right zero divisors. If $a \cdot b = 0$ with both a and b nonzero, one could call a a left zero divisor and b a right zero divisor, but we will not worry about such things. The name zero divisor comes from the usual meaning of divisor in $\mathbb{Z}$. If c and d are integers, then c is called a divisor of d if there is an integer e with $ce = d$. If $ce = 0$, then this terminology would lead us to say that c is a divisor of 0. However, since $c \cdot 0 = 0$ for all c, this would seem to lead us to call every integer a divisor of zero. This is not a useful statement. The restriction in the definition above to require $b \neq 0$ in order to call a a zero divisor if $ab = 0$ eliminates this worry.

Example 3.19. The rings $\mathbb{Z}$, $\mathbb{Q}$, $\mathbb{R}$, and $\mathbb{C}$ have no zero divisors. Each of these rings has the familiar property that $ab = 0$ implies $a = 0$ or $b = 0$.

Example 3.20. The ring of 2×2 matrices with real number entries has zero divisors, as the example above shows. In fact, if

$$A = \begin{pmatrix} a & b \\ c & d \end{pmatrix},$$

then the *adjoint* of A is the matrix

$$\mathrm{adj}(A) = \begin{pmatrix} d & -b \\ -c & a \end{pmatrix},$$

and from a simple calculation or a recollection from linear algebra,

$$A \cdot \mathrm{adj}(A) = \det(A) \begin{pmatrix} 1 & 0 \\ 0 & 1 \end{pmatrix} = \begin{pmatrix} ad - bc & 0 \\ 0 & ad - bc \end{pmatrix}.$$

Thus, if $\det(A) = 0$, then A is a zero divisor.

Example 3.21. We have seen that $\mathbb{Z}_6$ has zero divisors; for example, $\overline{2} \cdot \overline{3} = \overline{0}$. Similarly, $\mathbb{Z}_9$ has zero divisors, since $\overline{3} \cdot \overline{3} = \overline{0}$. Also, $\mathbb{Z}_{12}$ has zero divisors since $\overline{6} \cdot \overline{4} = \overline{24} = \overline{0}$. In contrast, $\mathbb{Z}_5$ has no zero divisors; if we view the multiplication table for $\mathbb{Z}_5$, we see that the product of two nonzero elements is always nonzero:

$\cdot$ mod 5	$\overline{0}$	$\overline{1}$	$\overline{2}$	$\overline{3}$	$\overline{4}$
$\overline{0}$	$\overline{0}$	$\overline{0}$	$\overline{0}$	$\overline{0}$	$\overline{0}$
$\overline{1}$	$\overline{0}$	$\overline{1}$	$\overline{2}$	$\overline{3}$	$\overline{4}$
$\overline{2}$	$\overline{0}$	$\overline{2}$	$\overline{4}$	$\overline{1}$	$\overline{3}$
$\overline{3}$	$\overline{0}$	$\overline{3}$	$\overline{1}$	$\overline{4}$	$\overline{2}$
$\overline{4}$	$\overline{0}$	$\overline{4}$	$\overline{3}$	$\overline{2}$	$\overline{1}$

Example 3.22. Let R and S be rings, and consider the ring $R \times S$ with operations defined earlier. Then this ring has zero divisors, as long as both R and S contain nonzero elements. For, if $r \in R$ and $s \in S$, then $(r, 0) \cdot (0, s) = (0, 0)$. For example, if $R = S = \mathbb{Z}$, then elements of the form $(n, 0)$ or $(0, m)$ with n, m nonzero are zero divisors in $\mathbb{Z} \times \mathbb{Z}$.

There is a cancellation law of multiplication for the ordinary number systems. Because of the possible presence of zero divisors, the generalization to arbitrary rings has to be phrased more carefully.

Proposition 3.23 (Cancellation Law of Multiplication). *Let R be a ring. Suppose that $a, b, c \in R$ with $ab = ac$ and $a \neq 0$. If a is not a zero divisor, then $b = c$.*

Proof. Let $a, b, c \in R$ with $ab = ac$. Suppose that a is not a zero divisor. From the equation we get $ab - ac = 0$, so $a(b - c) = 0$. Since a is not a zero divisor, $b - c = 0$, or $b = c$. □

The cancellation law fails if a is a zero divisor. For example, in $\mathbb{Z}_6$ we have $\overline{2} \cdot \overline{4} = \overline{2} \cdot \overline{1}$ even though $\overline{4} \neq \overline{1}$. Therefore, we cannot cancel $\overline{2}$ in such an equation. Similarly, with matrices we have the equation

$$\begin{pmatrix} 1 & 0 \\ 0 & 0 \end{pmatrix} \begin{pmatrix} 2 & 0 \\ -3 & 4 \end{pmatrix} = \begin{pmatrix} 1 & 0 \\ 0 & 0 \end{pmatrix} \begin{pmatrix} 2 & 0 \\ -3 & 6 \end{pmatrix}$$

while

$$\begin{pmatrix} 2 & 0 \\ -3 & 4 \end{pmatrix} \neq \begin{pmatrix} 2 & 0 \\ -3 & 6 \end{pmatrix}.$$

More generally, in any ring, if $ab = 0$ with $a, b \neq 0$, then $a \cdot b = a \cdot 0$, so cancellation of the zero divisor a is not valid.

We now return to the idea of multiplicative inverses. Recall the multiplication table for $\mathbb{Z}_6$:

$\cdot$ mod 6	$\overline{0}$	$\overline{1}$	$\overline{2}$	$\overline{3}$	$\overline{4}$	$\overline{5}$
$\overline{0}$	$\overline{0}$	$\overline{0}$	$\overline{0}$	$\overline{0}$	$\overline{0}$	$\overline{0}$
$\overline{1}$	$\overline{0}$	$\overline{1}$	$\overline{2}$	$\overline{3}$	$\overline{4}$	$\overline{5}$
$\overline{2}$	$\overline{0}$	$\overline{2}$	$\overline{4}$	$\overline{0}$	$\overline{2}$	$\overline{4}$
$\overline{3}$	$\overline{0}$	$\overline{3}$	$\overline{0}$	$\overline{3}$	$\overline{0}$	$\overline{3}$
$\overline{4}$	$\overline{0}$	$\overline{4}$	$\overline{2}$	$\overline{0}$	$\overline{4}$	$\overline{2}$
$\overline{5}$	$\overline{0}$	$\overline{5}$	$\overline{4}$	$\overline{3}$	$\overline{2}$	$\overline{1}$

From this table we see that $\overline{2}$, $\overline{3}$, and $\overline{4}$ are zero divisors. The remaining nonzero elements $\overline{1}$ and $\overline{5}$ are not. In fact, $\overline{1}$ and $\overline{5}$ each have a multiplicative inverse, namely itself, since $\overline{1} \cdot \overline{1} = \overline{1}$ and $\overline{5} \cdot \overline{5} = \overline{1}$.

Definition 3.24. If R is a ring, then an element $a \in R$ is said to be a unit if there is an element $b \in R$ with $a \cdot b = b \cdot a = 1$. In this case a and b are said to be multiplicative inverses.

A unit is never a zero divisor, as we now prove.

Proposition 3.25. *Let R be a ring. If $a \in R$ is a unit, then a is not a zero divisor.*

Proof. Let $a \in R$ be a unit with multiplicative inverse $c \in R$. Then $ac = ca = 1$. Suppose that $b \in R$ with $ab = 0$. Then

$$\begin{aligned} 0 &= c \cdot 0 = c(ab) \\ &= (ca)b = 1 \cdot b = b. \end{aligned}$$

We have shown that if $ab = 0$, then $b = 0$. Thus, a is not a zero divisor. An obvious modification of the argument just given shows that if $ba = 0$, then $b = 0$ as well. □

The example of $\mathbb{Z}$, with no zero divisors but only ± 1 for units, shows that the converse of this proposition does not hold in general.

Exercises

1. Let R be a ring and a an element of R. Prove that the additive inverse of a is unique, i.e., if $a + b = a + b' = 0$, then $b = b'$.
2. Find all solutions in $\mathbb{Z}_8$ to the equation $x^2 - \overline{1} = \overline{0}$.
3. Find all solutions in $\mathbb{Z}_{60}$ to the equation $x^2 - \overline{1} = \overline{0}$. How many did you find? How does this compare with your prior experience of the number of roots a polynomial of degree n can have?
4. In this problem you will show that there are infinitely many real-valued 2×2 matrices satisfying the equation $X^3 - I = 0$.

   (a) To begin let

   $$A = \begin{pmatrix} 0 & -1 \\ 1 & -1 \end{pmatrix}$$

   and show that A satisfies the equation $X^3 = I$.

   (b) Next let B be any invertible 2×2 matrix and prove that BAB^{-1} also satisfies the equation $X^3 - I = 0$.

   (c) Finally, verify that

   $$\begin{pmatrix} 1 & b \\ 0 & 1 \end{pmatrix} \begin{pmatrix} 0 & -1 \\ 1 & -1 \end{pmatrix} \begin{pmatrix} 1 & b \\ 0 & 1 \end{pmatrix}^{-1} = \begin{pmatrix} b & -b^2 - 1 - b \\ 1 & -1 - b \end{pmatrix},$$

   and conclude that there are infinitely many distinct 2×2 matrices satisfying the equation $X^3 - I = 0$.

5. Let R be a commutative ring in which there are no zero divisors. Prove that the only solutions to $x^2 - 1 = 0$ in R are 1 and -1.
   (Hint: factor the left-hand side of the equation.)
6. Let R and S be rings. Referring to Example 3.12, prove that $R \times S$ satisfies the distributive properties.
7. If R is a ring, prove that $a(-b) = (-a)b = -(ab)$ for all $a, b \in R$.
8. If R is a ring, prove that $a(b - c) = ab - ac$ for all $a, b, c \in R$.
9. Prove that the set R of Example 3.11 with the operations

   $$A + B = (A - B) \cup (B - A) = (A \cup B) - (A \cap B),$$
   $$A \cdot B = A \cap B.$$

   is a ring. What is the multiplicative identity? Which elements have multiplicative inverses?
10. Prove that if a and b are units in the ring R, then ab is a unit.
    (Hint: A fact from linear algebra about invertible matrices may stimulate your thinking.)
11. Prove that $\overline{a} \in \mathbb{Z}_n$ is a zero divisor if and only if $\gcd(a, n) > 1$.
12. Prove that a matrix $A \in M_n(\mathbb{R})$ is a zero divisor if and only if $\det(A) = 0$. (Hint: Recall two things:

    (a) $AB = 0$ means that the columns of B solve the homogeneous linear system $AX = 0$, and
    (b) The determinant criterion for the existence of nontrivial solutions to $AX = 0$.)

13. Prove that the zero divisors of $\mathbb{Z} \times \mathbb{Z}$ are the elements of the form $(a, 0)$ or $(0, a)$ for any $a \in \mathbb{Z}$.
14. Give an example of a ring R and an element $a \in R$ that is neither a zero divisor nor a unit.
15. Let R be a ring in which $0 = 1$. Prove that $R = \{0\}$.
16. Prove that the multiplicative identity of a ring is unique.
17. If a and b are elements of a ring R, prove that $(-a) \cdot (-b) = ab$.
18. Prove that if a and b are elements of a ring, then $(a + b)^2 = a^2 + ab + ba + b^2$.
19. Let R be a ring in which $a^2 = a$ for all $a \in R$. Prove that $-a = a$. Then prove that R is commutative.
20. Let R be a ring. If $a \in R$, define powers of a inductively by $a^0 = 1$ and $a^{n+1} = a^n \cdot a$.

    (a) If n and m are positive integers, prove that $a^n \cdot a^m = a^{n+m}$.
    (b) Prove or disprove by finding a counterexample: If $a, b \in R$, then $(ab)^n = a^n \cdot b^n$. If this is not always true, determine conditions under which it is true.

21. Define a relation between elements of a commutative ring R by $a \sim b$ if there is a unit $u \in R$ with $b = au$. Prove that this relation is an equivalence relation. Determine the equivalence classes of this relation if $R = \mathbb{Z}$ and if $R = \mathbb{Z}_{12}$.
22. Is subtraction on $\mathbb{Z}$ an associative operation? Is it commutative? If it is, prove it. If not, give an example to demonstrate that it is not.
23. Let R be a ring. Suppose that S is a subset of R containing 1 satisfying (i) if $a, b \in S$, then $a - b \in S$, and (ii) if $a, b \in S$, then $ab \in S$.

    (a) Show that $0 \in S$.
    (b) Show that if $a \in S$, then $-a \in S$.
    (c) Show that if $a, b \in S$, then $a + b \in S$. From this and (ii), we can view $+$ and $\cdot$ as operations on S.
    (d) Conclude that all of the ring axioms hold for S from the ring axioms for R, from the hypotheses, and from (a)–(c).
    The ring S is said to be a *subring* of R.

24. Let $\mathbb{H}$ be the set of all 2×2 matrices with complex number entries of the form $\begin{pmatrix} a & -b \\ \overline{b} & \overline{a} \end{pmatrix}$. Prove that $\mathbb{H}$ satisfies the hypotheses of the previous problem, with $R = M_2(\mathbb{C})$, so that we can consider $\mathbb{H}$ to be a subring of $M_2(\mathbb{C})$. Moreover, prove that $\mathbb{H}$ is noncommutative, and that every nonzero element of $\mathbb{H}$ has a multiplicative inverse in $\mathbb{H}$.

    (The ring $\mathbb{H}$, discovered in 1843 by Hamilton, was the first example discovered of a noncommutative ring in which every nonzero element has a multiplicative inverse. This matrix theoretic description is not Hamilton's original description. He was interested in an algebraic method to describe rotations in $\mathbb{R}^3$ analogous to how multiplication by the complex number i rotates the complex plane (hence $\mathbb{R}^2$) by 90° in the counterclockwise direction. The ring $\mathbb{H}$ is now called the ring of Hamilton's quaternions.)
25. Let $\mathbb{H}$ be the ring of the previous problem. Let

$$I = \begin{pmatrix} i & 0 \\ 0 & -i \end{pmatrix}, J = \begin{pmatrix} 0 & -1 \\ 1 & 0 \end{pmatrix}, K = IJ = \begin{pmatrix} 0 & -i \\ -i & 0 \end{pmatrix}.$$

We write 1 for the identity matrix. Prove that $I^2 = J^2 = K^2 = -1$ and $K = -JI$. Also, note that if $a = \alpha + \beta i$ and $b = \gamma + \delta i$, then

$$\begin{pmatrix} a & -b \\ \overline{b} & \overline{a} \end{pmatrix} = \alpha \cdot 1 + \beta I + \gamma J + \delta K,$$

if we view 1 as the 2×2 identity matrix. Finally, if $\Delta = \alpha^2 + \beta^2 + \gamma^2 + \delta^2$, show that

$$(\alpha \cdot 1 + \beta I + \gamma J + \delta K)^{-1} = \left(\frac{\alpha}{\Delta}\right) \cdot 1 - \left(\frac{\beta}{\Delta}\right) I - \left(\frac{\gamma}{\Delta}\right) J - \left(\frac{\delta}{\Delta}\right) K.$$

Remark 3.26. Hamilton defined the quaternions as all symbols of the form $a + bi + cj + dk$ with $a, b, c, d \in \mathbb{R}$ with addition given by

$$(a + bi + cj + dk) + (a' + b'i + c'j + d'k) = (a + a') + (b + b')i + (c + c')j + (d + d')k$$

and multiplication obtained by using the distributive law along with the identities $i^2 = j^2 = k^2 = -1$ and $k = ij = -ji$.

3.3 Fields

In the previous section we discussed the notion of a zero divisor and saw that if an element has a multiplicative inverse then it is not a zero divisor. The ring $\mathbb{Z}$ has no zero divisors. However, few integers have multiplicative inverses in $\mathbb{Z}$; in fact, only 1 and -1 have multiplicative inverses in $\mathbb{Z}$. On the other hand, every nonzero integer has a multiplicative inverse in the ring $\mathbb{Q}$. Even more than that, every nonzero element of $\mathbb{Q}$ has a multiplicative inverse in $\mathbb{Q}$. A similar statement holds for $\mathbb{R}$ and for $\mathbb{C}$. Therefore, in $\mathbb{Q}$, $\mathbb{R}$, and $\mathbb{C}$ we may divide, as long as we do not try to divide by 0. However, if we restrict ourselves to $\mathbb{Z}$, we cannot always divide; we must extend to $\mathbb{Q}$ in order to do general division of integers. Rings in which we can divide are important enough, and common enough, for us to investigate.

Definition 3.27. A field is a commutative ring F containing at least two elements such that every nonzero element of F has a multiplicative inverse in F.

The hypothesis that a field contains at least two elements ensures that $0 \neq 1$; necessarily $0 = 1$ if the field has only one element. If it has more than one, a short argument will prove that $0 \neq 1$. This assumption is not important, but it is convenient. It is similar to the assumption that 1 is not a prime number, even though it cannot be factored in a nontrivial way.

Example 3.28. The number systems $\mathbb{Q}$, $\mathbb{R}$, and $\mathbb{C}$ are all examples of fields.

The following proposition gives us infinitely many examples of fields.

Proposition 3.29. *Let n be a positive integer. Then $\mathbb{Z}_n$ is a field if and only if n is prime.*

Proof. Let n be a positive integer. We know that $\mathbb{Z}_n$ is a commutative ring, so it suffices to prove that every nonzero element of $\mathbb{Z}_n$ has a multiplicative inverse if and only if n is prime. First, suppose that n is prime. Let $\overline{a} \in \mathbb{Z}_n$ be nonzero. Then $a \not\equiv 0 \bmod n$. Consequently, n does not divide a. Since n is prime, this forces $\gcd(a, n) = 1$. Thus, by Proposition 1.13, there are integers s and t with $1 = as + nt$. This means $as \equiv 1 \bmod n$, or $\overline{a} \cdot \overline{s} = \overline{1}$ in $\mathbb{Z}_n$. We have thus produced a multiplicative inverse for $\overline{a}$. Since $\overline{a}$ was an arbitrary nonzero element, this proves that $\mathbb{Z}_n$ is a field.

For the converse, we prove the contrapositive: If n is not prime, then $\mathbb{Z}_n$ is not a field. Assume that n is composite and factors as $n = ab$ with $1 < a, b < n$. Then $\overline{a} \cdot \overline{b} = \overline{ab} = \overline{n} = \overline{0}$. Since $\overline{a} \neq \overline{0}$, and $\overline{b} \neq \overline{0}$ we find that both are zero divisors. But a field has no zero divisors, so $\mathbb{Z}_n$ is not a field. □

Example 3.30. Let $F = \{0, 1, a, b\}$ be a set with four elements, and define operations by the following tables

$+$	0	1	a	b
0	0	1	a	b
1	1	0	b	a
a	a	b	0	1
b	b	a	1	0

and

$\cdot$	0	1	a	b
0	0	0	0	0
1	0	1	a	b
a	0	a	b	1
b	0	b	1	a

.

A tedious computation, which we will not bother with, shows that F, together with these operations, is a field. This example looks very ad-hoc. However, there is a systematic way to obtain this example from the field $\mathbb{Z}_2$ that is analogous to how $\mathbb{C}$ is obtained from $\mathbb{R}$ by attaching a square root of -1. This method of building fields from others is the basis for building Reed–Solomon codes, an important class of error correcting codes discussed in Chap. 7.

Example 3.31. Fields can be found among certain subsets of $\mathbb{R}$ or $\mathbb{C}$. For example, let

$$\mathbb{Q}(\sqrt{2}) = \left\{a + b\sqrt{2} : a, b \in \mathbb{Q}\right\}$$

and

$$\mathbb{Q}(i) = \{a + bi : a, b \in \mathbb{Q}\}$$

The set $\mathbb{Q}(\sqrt{2})$ is a subset of $\mathbb{R}$ and $\mathbb{Q}(i)$ is a subset of $\mathbb{C}$. We (partially) verify that $\mathbb{Q}(\sqrt{2})$ is a field and leave the other example for an exercise. We first note that ordinary addition and multiplication yield operations on $\mathbb{Q}(\sqrt{2})$. To see this, let $a + b\sqrt{2}$ and $c + d\sqrt{2}$ be elements of $\mathbb{Q}(\sqrt{2})$. Implicit in this statement is that $a, b, c, d \in \mathbb{Q}$. Then

$$(a + b\sqrt{2}) + (c + d\sqrt{2}) = (a + c) + (b + d)\sqrt{2}.$$

Since $a + c$ and $b + d$ are rational numbers, this sum lies in $\mathbb{Q}(\sqrt{2})$. Thus, the sum of two elements of $\mathbb{Q}(\sqrt{2})$ is an element of $\mathbb{Q}(\sqrt{2})$. For multiplication, we have

$$\begin{aligned}(a + b\sqrt{2}) \cdot (c + d\sqrt{2}) &= ac + ad\sqrt{2} + bc\sqrt{2} + 2bd \\ &= (ac + 2bd) + (ad + bc)\sqrt{2}.\end{aligned}$$

Since $ac + 2bd$ and $ad + bc$ are rational numbers, this product lies in $\mathbb{Q}(\sqrt{2})$. Therefore, we have operations of addition and multiplication on $\mathbb{Q}(\sqrt{2})$. Almost all of the properties to be a field are automatic. For example, commutativity of addition in $\mathbb{Q}(\sqrt{2})$ is a special case of commutativity of addition in $\mathbb{R}$. Similarly, since the additive identity 0 and multiplicative identity 1 in $\mathbb{R}$ can be written, respectively, as $0 = 0 + 0\sqrt{2}$ and $1 = 1 + 0\sqrt{2}$, and hence are elements of $\mathbb{Q}(\sqrt{2})$, we see that $\mathbb{Q}(\sqrt{2})$ does contain the multiplicative and additive identities. Also, if $a, b \in \mathbb{Q}$, then the additive inverse of $a + b\sqrt{2}$ is $-(a + b\sqrt{2}) = (-a) + (-b)\sqrt{2}$, which lies in $\mathbb{Q}(\sqrt{2})$. Finally, we check for multiplicative inverses. This is the most involved part of verifying that $\mathbb{Q}(\sqrt{2})$ is a field. Let $a, b \in \mathbb{Q}$, and consider a nonzero element $a + b\sqrt{2}$. In order for this number to be nonzero, either a or b is nonzero. By rationalizing the denominator, similar to the trick of multiplying by the conjugate of a complex number, we have

$$\frac{1}{a+b\sqrt{2}} = \frac{a-b\sqrt{2}}{(a-b\sqrt{2})(a+b\sqrt{2})} = \frac{a-b\sqrt{2}}{a^2-2b^2}$$
$$= \left(\frac{a}{a^2-2b^2}\right) + \left(\frac{-b}{a^2-2b^2}\right)\sqrt{2}.$$

This calculation makes sense, and furthermore shows that $a+b\sqrt{2}$ has a multiplicative inverse in $\mathbb{Q}(\sqrt{2})$, provided of course that $a^2-2b^2 \neq 0$. In that case the last expression can be viewed as a linear combination of 1 and $\sqrt{2}$ with rational number coefficients. It remains to verify that $a^2-2b^2 \neq 0$ always holds when either a or b is nonzero. To that end, assume that at least one of a and b is nonzero and view the calculation $(a+b\sqrt{2})(a-b\sqrt{2}) = a^2-2b^2$ as an equation in $\mathbb{R}$, which has no zero divisors. By assumption $a+b\sqrt{2} \neq 0$ so if $a-b\sqrt{2}$ is also nonzero, then $a^2-2b^2 \neq 0$ since it is a product of two nonzero real numbers. If $a-b\sqrt{2}=0$ and $b=0$, then $a=0$, but we know that either a or b is nonzero. If $b \neq 0$, then $a-b\sqrt{2}=0$ yields $\sqrt{2}=a/b$, a rational number. However, $\sqrt{2}$ is irrational (Exercise 7 below). Thus $a-b\sqrt{2} \neq 0$, so that $a^2-2b^2 \neq 0$, and multiplicative inverses for nonzero elements in $\mathbb{Q}(\sqrt{2})$ always exist, i.e., $\mathbb{Q}(\sqrt{2})$ is a field.

It is not at all apparent that Examples 3.30 and 3.31 are related in any way. Later on, when we discuss the method to build fields from others by adjoining roots of polynomials hinted at in the previous example (i.e., the polynomial x^2-2 with coefficients in $\mathbb{Q}$ has two roots in $\mathbb{Q}(\sqrt{2})$), we will see that these two examples can actually be obtained by the same method applied to different base fields. The former example is built from $\mathbb{Z}_2$ and the latter from $\mathbb{Q}$.

Definition 3.32. A commutative ring without zero divisors is called an integral domain.

Example 3.33. As the name implies, the ring of integers is an integral domain. Every field is an integral domain, as is $\mathbb{R}[x]$ the ring of all polynomials with real number coefficients.

The proof of Proposition 3.29 shows that the ring $\mathbb{Z}_n$ is an integral domain if and only if n is prime. This proposition has the following generalization.

Theorem 3.34. *Every finite integral domain is a field.*

Proof. Let D be a finite integral domain and $S = \{d_1, d_2, \ldots, d_n\}$ the set of its nonzero elements with $d_i \neq d_j$ for $i \neq j$. We must show that each d_i has a multiplicative inverse in S. To that end, denote by d_iS the set $\{d_id_1, d_id_2, \ldots, d_id_n\}$ and observe that $d_iS \subseteq S$ because D is an integral domain. But in fact $d_iS = S$, since otherwise there would be repetition among the elements d_id_j. However $d_id_j = d_id_k$ holds only for $j = k$ by the cancellation law for multiplication. Thus $d_iS = S$ and, because $1 \in S$, it must be the case that $1 = d_id_j$ for some d_j. □

Exercises

1. Recall that by reversing the steps in the Euclidean Algorithm one can always express the greatest common divisor of a pair of integers as an integer linear combination of them: $\gcd(a,b) = ax+by$ for certain integers x, y. Use this to find an algorithm to determine the multiplicative inverse of an arbitrary nonzero element $\overline{a}$ in the field $\mathbb{Z}_p$ (p prime).
2. Let F be a finite field. Prove that there is a positive integer n such that $\underbrace{1+1+\cdots+1}_{n \text{ times}} = 0$. Prove that the least such n is a prime number.
3. Let F be a field. Prove that every linear equation $ax+b=0$ with $a, b \in F$ has a unique solution in F.

4. Find a quadratic equation with coefficients in $\mathbb{Z}_2$ that has no solution in $\mathbb{Z}_2$.
5. Find a quadratic equation with coefficients in $\mathbb{Z}_3$ that has no solution in $\mathbb{Z}_3$.
6. Find a cubic equation with coefficients in $\mathbb{Z}_2$ that has no solution in $\mathbb{Z}_2$.
7. Use unique factorization in $\mathbb{Z}$ to prove that $\sqrt{2}$ is irrational (i.e., assume that $\sqrt{2} = \frac{a}{b}$ with relatively prime positive integers a and b. From $2b^2 = a^2$ deduce that a and b are both even).
8. Use unique factorization in $\mathbb{Z}$ to prove that $\sqrt[3]{2}$ is irrational.

References

1. Richman F (1971) Number theory, an introduction to algebra. Brooks-Cole, Monterey
2. van der Waerden BL (1971) Modern algebra. Springer, Berlin

Chapter 4
Linear Algebra and Linear Codes

In this chapter we review the main ideas of linear algebra. The one twist is that we allow our scalars to come from any field instead of just the field of real numbers. In particular, the notion of a vector space over the field $\mathbb{Z}_2$ will be essential in our study of coding theory. We will also need to look at other finite fields when we discuss Reed–Solomon codes. One benefit to working with finite dimensional vector spaces over finite fields is that all sets in question are finite, and so computers can be useful in working with them.

4.1 Vector Spaces

Let us review the idea of a vector space. Recall that two types of objects, *vectors* and *scalars* are involved. For example, $\mathbb{R}^3$, the usual three-dimensional space, can be interpreted as a set of vectors if we take as the set of scalars $\mathbb{R}$, the set of real numbers. While vector spaces with the real number field as the set of scalars may have been the only context in your previous study of linear algebra, it is true that in any computation or definition about vector spaces, all you needed was the ability to add, subtract, multiply, and divide scalars, and that these arithmetic operations had some appropriate properties. In fact, there is nothing we need about $\mathbb{R}$ other than its properties as a field to do everything you have seen about vector spaces. Therefore, we will simply review these concepts, assuming only that the set of scalars is some field F instead of $\mathbb{R}$.

In a vector space V we have two operations, (vector) addition and scalar multiplication. The addition operation is a binary operation on V. Scalar multiplication is somewhat different in that the input is a pair consisting of an element of the scalar field F and a vector in V, and the output is a vector, so scalar multiplication is really a function from $F \times V$ to V. In keeping with convention, write this function as $\alpha \cdot v$ or simply αv rather than $\cdot(\alpha, v)$.

Definition 4.1. Let F be a field. An F-vector space is a nonempty set V together with an operation $+$ on V and an operation $\cdot : F \times V \to V$, with the following properties (here u, v, w denote arbitrary elements of V, and α, β arbitrary elements of F):

- $u + v$ and $\alpha \cdot u$ are elements of V;
- $u + v = v + u$;
- $u + (v + w) = (u + v) + w$;
- There is a vector $\mathbf{0}$ with $u + \mathbf{0} = u$ for any $u \in V$;
- For every $u \in V$ there is a vector $-u \in V$ with $u + (-u) = \mathbf{0}$;

D.R. Finston and P.J. Morandi, *Abstract Algebra: Structure and Application*,
Springer Undergraduate Texts in Mathematics and Technology, DOI 10.1007/978-3-319-04498-9_4

- $\alpha(u + v) = \alpha u + \alpha v$;
- $(\alpha + \beta)u = \alpha u + \beta u$;
- $(\alpha\beta)u = \alpha(\beta u)$;
- $1 \cdot u = u$.

Naturally $+$ is referred to as addition in the vector space and $\cdot$ as scalar multiplication. Before looking at examples, we point out that much of our study of abstract algebra involves an analysis of structures and their operations. The structure of a vector space comes with an addition whose properties are exactly the same as for addition in the definition of a ring. Ring multiplication is absent, but we have scalar multiplication that intertwines the addition with the multiplication in the base field. Later on we will focus on structures with only one operation. As with the definition of rings, the above list of vector space properties begins with closure statements, i.e., that addition and scalar multiplication are operations.

We now illustrate the definition of a vector space with several examples.

Example 4.2. The most basic example we will use of an F-vector space is the space F^n of all n-tuples of elements of F. That is,

$$F^n = \{(a_1, \ldots, a_n) : a_i \in F\}.$$

The operations that make F^n into a vector space are the pointwise operations coming from the addition and multiplication in the field F:

$$(a_1, \ldots, a_n) + (b_1, \ldots, b_n) = (a_1 + b_1, \ldots, a_n + b_n),$$
$$\alpha(a_1, \ldots, a_n) = (\alpha a_1, \ldots, \alpha a_n).$$

Example 4.3. Let V be the set of all $n \times m$ matrices with real number entries. We will denote by (a_{ij}) the matrix whose i, j entry is a_{ij}. Recall that matrix addition and scalar multiplication are defined by

$$(a_{ij}) + (b_{ij}) = (a_{ij} + b_{ij}),$$
$$\alpha(a_{ij}) = (\alpha a_{ij}).$$

In other words, add and do scalar multiplication componentwise. You should refer to a linear algebra text if necessary to recall that matrix arithmetic satisfies the properties required of an $\mathbb{R}$-vector space.

Example 4.4. We generalize the previous example a little. Let us replace $\mathbb{R}$ by any field F, and make V to be the set of all $n \times m$ matrices with entries from F. We use the same formulas to define addition and scalar multiplication. The proof that shows the set of real-valued matrices of a given size forms a vector space also shows, without any changes, that the set of matrices with entries in F also is a vector space.

Example 4.5. Let F be a field. We denote by $F[x]$ the set of polynomials with coefficients in F. Our usual definitions of addition and multiplication in $\mathbb{R}[x]$ work fine for $F[x]$. As a special example of multiplication, we can define scalar multiplication by $\alpha \sum_{i=0}^{n} a_i x^i = \sum_{i=0}^{n} \alpha a_i x^i$. With the operations of addition and scalar multiplication, a routine argument will show that $F[x]$ is an F-vector space. In fact, $F[x]$, together with addition and multiplication, is also a commutative ring. Thus, $F[x]$ gives us an example of both a ring and a vector space as does the ring of $n \times n$. matrices over the field F; the former is a commutative ring while the latter is noncommutative if $n \geq 2$. A structure of this type is called a commutative F-algebra if the ring structure is commutative and a noncomutative F-algebra otherwise.

We now prove two simple properties of vector spaces. These are similar in spirit to statements we proved for rings. First of all, note that the proofs of uniqueness of the additive identity and the additive inverse of an element of a ring work word for word for vector spaces. This is because the proofs only involved addition and, with respect to those axioms dealing only with their additive structures, there is no difference between a ring and a vector space.

Lemma 4.6. *Let F be a field and let V be an F-vector space. If $v \in V$, then $0 \cdot v = \mathbf{0}$.*

Proof. Let $v \in V$. To show that $0 \cdot v = \mathbf{0}$, we must show that $0 \cdot v$ satisfies the definition of the additive identity. By using one of the distributive laws, we see that

$$\begin{aligned} 0 \cdot v + v &= 0 \cdot v + 1 \cdot v = (0 + 1) \cdot v \\ &= 1 \cdot v = v = \mathbf{0} + v. \end{aligned}$$

By the cancellation law of addition, which holds for vector spaces just as for rings, we get $0 \cdot v = \mathbf{0}$. □

Lemma 4.7. *Let F be a field and let V be an F-vector space. If $v \in V$, then $(-1) \cdot v = -v$.*

Proof. Let $v \in V$. To prove that $(-1) \cdot v$ is equal to $-v$, we need to prove that it is the additive inverse of v. We have

$$\begin{aligned} v + (-1) \cdot v &= 1 \cdot v + (-1) \cdot v = (1 + (-1)) \cdot v \\ &= 0 \cdot v = \mathbf{0} \end{aligned}$$

by the previous lemma. These equations do indeed show that $(-1) \cdot v$ is the additive inverse of v. □

To consider the types of examples that most often arise, we need the notion of a subspace.

Definition 4.8. Let F be a field and let V be an F-vector space. A nonempty subset W of V is called a subspace of V if the operations on V induce operations on W, and if W is an F-vector space under these operations.

Let us explain this definition. Addition on V is a function from $V \times V$ to V. The product $W \times W$ is a subset of $V \times V$, so, with addition viewed as a function $V \times V \to V$, we can restrict it to this smaller domain. In general, the image of this function need not be inside W. However, if it is, i.e., if addition restricts to a function $W \times W$ to W, then the restriction is a binary operation on W. When this occurs, we say that W is *closed under the addition on V*. To say this more symbolically, to say that W is closed under addition means that for all $v, w \in W$, the sum $v + w$ must also be an element of W. Similarly, the scalar multiplication operation on V, which is a function from $F \times V$ to V, may restrict to a function $F \times W$ to W. If this happens, we say that W is closed under scalar multiplication. Symbolically, this means that for all $\alpha \in F$ and all $v \in W$, the element αv must be an element of W. When W is closed under addition and scalar multiplication, then we have appropriate operations on W to discuss whether or not W is a vector space. In fact, as the following lemma shows, if W is closed under both addition and scalar multiplication, then W is automatically a subspace of V.

Lemma 4.9. *Let F be a field and let V be an F-vector space. Let W be a nonempty subset of V, and suppose that (i) for each $v, w \in W$, the sum $v + w \in W$; and (ii) for each $\alpha \in F$ and $v \in W$, the product $\alpha v \in W$. Then W is a subspace of V.*

Proof. Suppose W satisfies statements (i) and (ii) in the lemma. We must verify that W, together with its induced operations, satisfies the definition of a vector space. Thus, we must verify the eight properties of the definition. Six of these properties are immediate from the fact that V is a vector space. For instance, the commutative law of addition for V yields $v + w = w + v$ for all $v, w \in V$. In particular, if $v, w \in W$, then v and w are also in V, so $v + w = w + v$. In a similar way, the associative property, the two distributive properties, and the property $\alpha(\beta v) = (\alpha\beta)v$ for $\alpha, \beta \in F$ and $v \in W$ all hold. We therefore need to check the two remaining properties. First, we need the existence of an additive identity. It is sufficient to show that if $\mathbf{0}$ is the identity of V, then $\mathbf{0} \in W$. However, we are assuming that W is nonempty; suppose $w \in W$ is any element of W. Since W is closed under scalar multiplication, if $\alpha \in F$ is any scalar, then $\alpha w \in W$. In particular, choose $\alpha = 0$ so that $\mathbf{0} = 0w \in W$ as desired. Finally, we need to have the additive inverse of each element of W to be inside W. Let $w \in W$. Again, by closure of scalar multiplication, we see that $(-1) \cdot w \in W$. But, since $(-1) \cdot w = -w$, the additive inverse of w lies in W. This completes the proof. □

Example 4.10. Let $u_1 = (1, 2, 3)$ and $u_2 = (4, 5, 6)$, two elements of the $\mathbb{R}$-vector space $\mathbb{R}^3$. Let W be the set

$$W = \{\alpha u_1 + \beta u_2 : \alpha, \beta \in \mathbb{R}\}.$$

This is the set of all *linear combinations* of u_1 and u_2, or the span of u_1 and u_2, written $\text{Span}\{u_1, u_2\}$. We claim that W is a subspace of $\mathbb{R}^3$. Note that W is nonempty, since both u_1 and u_2 are in W. That u_1 is in W is because $u_1 = 1 \cdot u_1 + 0 \cdot u_2$ and similarly for u_2. To see that W is a subspace, we then need to verify that W is closed under addition and scalar multiplication. For addition, let $v, w \in W$. By definition of W, there are scalars $\alpha, \beta, \gamma, \delta$ with $v = \alpha u_1 + \beta u_2$ and $w = \gamma u_1 + \delta u_2$. Then

$$\begin{aligned} v + w &= (\alpha u_1 + \beta u_2) + (\gamma u_1 + \delta u_2) \\ &= (\alpha + \gamma)u_1 + (\beta + \delta)u_2. \end{aligned}$$

Since $\alpha + \gamma$ and $\beta + \delta$ are real numbers, the vector $v + w$ has the form necessary to show that it is in element of W. Therefore, W is closed under addition. Next, for scalar multiplication, let $v \in W$ and let c be a scalar. Again, we may write $v = \alpha u_1 + \beta u_2$ for some scalars α, β. Then

$$\begin{aligned} cv &= c(\alpha u_1 + \beta u_2) = c(\alpha u_1) + c(\beta u_2) \\ &= (c\alpha)u_1 + (c\beta)u_2, \end{aligned}$$

which shows that cv is an element of W. Therefore, by Lemma 4.9, W is a subspace of $\mathbb{R}^3$. Notice that we only used that u_1 and u_2 are vectors in a vector space, and not that the vector space is $\mathbb{R}^3$ and that u_1 and u_2 are prescribed. This will allow us to easily generalize this example in an important way.

We can be a little more precise about what is W. Since $v = (1, 2, 3)$ and $w = (4, 5, 6)$, we have

$$\begin{aligned} W &= \{\alpha(1, 2, 3) + \beta(4, 5, 6) : \alpha, \beta \in \mathbb{R}\} \\ &= \{(\alpha + 4\beta, 2\alpha + 5\beta, 3\alpha + 6\beta : \alpha, \beta \in \mathbb{R}\}. \end{aligned}$$

In other words, W consists of all three-tuples that can be obtained as the product

$$\begin{pmatrix} 1 & 4 \\ 2 & 5 \\ 3 & 6 \end{pmatrix} \begin{pmatrix} \alpha \\ \beta \end{pmatrix}$$

for any $\alpha, \beta \in \mathbb{R}$. This observation was already used in the study of linear codes and will continue to be important for the further study of coding theory. If you recall your multivariable calculus, you can see a more geometric description of W in terms of the cross product. Calculating $v \times w$ yields the vector $(-3, 6, -3)$ which is perpendicular to both v and w. Thus, using the dot product, we have $v \cdot (-3, 6, -3) = \mathbf{0}$ and $w \cdot (-3, 6, -3) = \mathbf{0}$. In fact, some computation shows that W is exactly the set of vectors that satisfy the equation $x \cdot (-3, 6, -3) = \mathbf{0}$. Geometrically, this set is a plane that passes through the origin.

Example 4.11. Let

$$A = \begin{pmatrix} 1 & 2 & 3 \\ 4 & 5 & 6 \\ 7 & 8 & 9 \end{pmatrix}.$$

Consider the *nullspace* (or *kernel*) of A, i.e., the set W of all elements $\mathbf{x} = (\alpha, \beta, \gamma)$ of $\mathbb{R}^3$ that satisfy $A\mathbf{x} = \mathbf{0}$. We claim that W is a subspace of $\mathbb{R}^3$. To do this we will use two properties of matrix multiplication. First, note that W is nonempty since the zero vector $(0, 0, 0)$ satisfies the equation $A\mathbf{x} = \mathbf{0}$. Next, let $v, w \in W$. Then $Av = \mathbf{0}$ and $Aw = \mathbf{0}$. Thus, $A(v + w) = Av + Aw = \mathbf{0} + \mathbf{0} = \mathbf{0}$, so $v + w \in W$. Finally, if $v \in W$ and $\alpha \in \mathbb{R}$, then $A(\alpha v) = \alpha(Av) = \alpha\mathbf{0} = \mathbf{0}$. Thus, $\alpha v \in W$, and so W is a subspace of $\mathbb{R}^3$. As with the previous example, we did not use that A was a specific 3×3 matrix, only that A was a matrix.

The properties of matrix multiplication we used in the proof above are (i) $A(B + C) = AB + AC$, and (ii) $A(\alpha B) = \alpha(AB)$ for any matrices A, B, C and any scalar C. These facts can be found in any linear algebra text.

Exercises

1. Let w be a fixed vector in $\mathbb{R}^3$, and let $W = \{v \in \mathbb{R}^3 : w \cdot v = 0\}$. Show that W is a subspace of $\mathbb{R}^3$. Find a matrix A so that W is the nullspace of A.
2. Let n be a positive integer, and let W be the set of all words in $\mathbb{Z}_2^n$ of even weight. Show that W is a subspace of $\mathbb{Z}_2^n$.
3. Let V be a vector space over a field F. If $v_1, \dots, v_n \in V$, show that the set $W = \{a_1 v_1 + \cdots + a_n v_n : a_i \in F\}$ is a subspace of V. This subspace is called the *span* of $\{v_1, \dots, v_n\}$.
4. Let $\mathrm{M_n}(F)$ be the set of all $n \times n$ matrices with entries from the field F.

   a. Is the set of all upper triangular $n \times n$ matrices a subspace of $\mathrm{M_n}(F)$? (Prove or explain why not.)
   b. Is the set of all nonsingular $n \times n$ matrices a subspace of $\mathrm{M_n}(F)$? (Prove or explain why not.)

5. Verify that Hamilton's quaternions, the ring $\mathbb{H}$ of Exercises 24 and 25 of Sect. 3.2, is a noncommutative $\mathbb{R}$-algebra. Find a subspace of $\mathbb{H}$ which is a commutative $\mathbb{R}$-algebra that can be identified with the field $\mathbb{C}$.

4.2 Linear Independence, Spanning, and Bases

In the previous section we defined vector spaces over an arbitrary field, gave some examples, and proved some simple properties about them. In this section we discuss the most important notions associated with vector spaces, those that lead to the idea of a basis. From both a theoretical and computational point of view, a basis is extremely important. Using a basis simplifies many computations and also aids in the proofs of many results. As we will see, this definition and other familiar concepts, structures, and operations from vector spaces over the real numbers carry over naturally to our more general setting.

One primary notion that has arisen earlier is that of a linear combination. To formalize it we make the following definition:

Definition 4.12. Let F be a field and let V be an F-vector space. Given a finite collection $\{v_1, v_2, \ldots, v_n\}$ vectors in V and a finite collection $\{a_1, a_2, \ldots, a_n\}$ of scalars in F, the vector $\sum_{i=1}^{n} a_i v_i$ is called a *linear combination* of the v_i.

Note the finiteness implicit in the definition. One may speak of a linear combination of infinitely many vectors, but the understanding is that all but finitely many of the scalar coefficients are equal to zero.

The next crucial notion is that of linear independence.

Definition 4.13. Let F be a field and let V be an F-vector space. A collection $\{v_1, \ldots, v_n\}$ of elements of V is said to be linearly independent if whenever there are scalars $a_1, \ldots, a_n$ with $\sum_{i=1}^{n} a_i v_i = \mathbf{0}$, then each $a_i = 0$. The set is linearly dependent if it is not linearly independent. More generally, an arbitrary subset S of V is linearly independent provided every linear combination of vectors in S resulting in the zero vector has all scalar coefficients equal to 0.

Example 4.14. Let F be a field and $F[x]$ the ring of polynomials in the variable x. Then the set $S = \{1, x, x^2, \ldots\}$ is linearly independent, since a linear combination $\sum_{i=1}^{n} a_i x^{m_i}$ of elements of S is a polynomial in x. It is the zero polynomial if and only of all the coefficients a_i are equal to 0.

Remark 4.15. While the definition given of linear independence is standard, technically it is problematic. For example, if v is any nonzero vector in V, then certainly $\{v\}$ is linearly independent. However, as sets, $\{v\} = \{v, v\}$. So if $v_1 = v_2 = v$, then the linear combination $1 \cdot v_1 + (-1) \cdot v_2 = v - v = 0$ says that $\{v\} = \{v_1, v_2\}$ is linearly dependent. The situation can be repaired if, instead of sets of vectors, we use sequences or ordered lists of vectors. The authors feel that this bit of logical fastidiousness doesn't justify a departure from standard terminology.

Definition 4.16. Let F be a field and let V be an F-vector space. A collection $\{v_1, \ldots, v_n\}$ of elements of V is said to be a spanning set for V if every element of V can be expressed in the form $\sum_{i=1}^{n} a_i v_i$ for some choice of scalars $a_1, \ldots, a_n$.

These two definitions together give us the notion of a basis.

Definition 4.17. Let F be a field and let V be an F-vector space. A subset $\{v_1, \ldots, v_n\}$ of V is said to be a basis for V if it is both linearly independent and a spanning set for V.

To help understand these definitions, we look at several examples.

Example 4.18. Let $i = (1, 0, 0)$, $j = (0, 1, 0)$, and $k = (0, 0, 1)$, three elements of the $\mathbb{R}$-vector space $\mathbb{R}^3$. Then $\{i, j, k\}$ is a basis. To verify this we need to prove that this set is linearly independent and spans $\mathbb{R}^3$. First, for independence, suppose there are scalars a, b, and c with $ai + bj + ck = \mathbf{0}$. Then

$$\begin{aligned}(0,0,0) &= ai + bj + ck = a(1,0,0) + b(0,1,0) + c(0,0,1) \\ &= (a,0,0) + (0,b,0) + (0,0,c) = (a,b,c).\end{aligned}$$

This vector equation yields $a = b = c = 0$. Thus, $\{i, j, k\}$ is indeed linearly independent. For spanning, if (x, y, z) is a vector in $\mathbb{R}^3$, then the equation $ai + bj + ck = (a, b, c)$ above shows us that we can write this vector in terms of i, j, and k as $(x, y, z) = xi + yj + zk$. Thus, any vector in $\mathbb{R}^3$ is a linear combination of $\{i, j, k\}$, so this set spans $\mathbb{R}^3$. Since this set both spans $\mathbb{R}^3$ and is independent, it is a basis of $\mathbb{R}^3$.

Example 4.19. Let $W = \{(a, b, 0) : a, b \in \mathbb{R}\}$, a subset of $\mathbb{R}^3$. In fact, W is a subspace of $\mathbb{R}^3$. The vectors $i = (1, 0, 0)$ and $j = (0, 1, 0)$ are elements of W. We claim that $\{i, j\}$ is a basis for W. An argument similar to that of the previous example will verify this; to summarize the ideas, we have $(a, b, 0) = ai + bj$ is the unique way of expressing the vector $(a, b, 0)$ as a linear combination of i and j. This proves spanning, and the uniqueness will show linear independence. Indeed $0i + 0j = (0, 0, 0)$, so the uniqueness shows that if $xi + yj = (0, 0, 0)$, then $x = y = 0$.

Example 4.20. Recall the Hamming code C of Chap. 2. It is the nullspace of

$$H = \begin{pmatrix} 0\,0\,0\,1\,1\,1\,1 \\ 0\,1\,1\,0\,0\,1\,1 \\ 1\,0\,1\,0\,1\,0\,1 \end{pmatrix},$$

i.e., the set of solutions to the matrix equation $Hx = \mathbf{0}$. C is a subspace of $\mathbb{Z}_2^7$; to verify this, first suppose that $v, w \in C$. Then $Hv = Hw = \mathbf{0}$. Thus, $H(v + w) = Hv + Hw = \mathbf{0} + \mathbf{0} = \mathbf{0}$, so $v + w \in C$. Also, if $v \in C$ and $\alpha \in \mathbb{Z}_2$, then $H(\alpha v) = \alpha(Hv) = \alpha\mathbf{0} = \mathbf{0}$. Thus, C is closed under addition and scalar multiplication, so it is indeed a subspace of $\mathbb{Z}_2^7$. We have seen that an arbitrary solution $(x_1, x_2, x_3, x_4, x_5, x_6, x_7)$ to $Hx = \mathbf{0}$ satisfies

$$\begin{aligned} x_1 &= x_3 + x_5 + x_7, \\ x_2 &= x_3 + x_6 + x_7, \\ x_4 &= x_5 + x_6 + x_7, \text{ and} \\ &x_3, x_5, x_6, x_7 \text{ are arbitrary.} \end{aligned}$$

We get four solutions by setting one arbitrary variable equal to 1 and the other three to 0. Doing so yields

$$\begin{aligned} c_1 &= (1,1,1,0,0,0,0), \; c_2 = (1,0,0,1,1,0,0), \\ c_3 &= (0,1,0,1,0,1,0), \; c_4 = (1,1,0,1,0,0,1). \end{aligned}$$

We claim that $\{c_1, c_2, c_3, c_4\}$ constitutes a basis for C. This is seen by noting that

$$\begin{pmatrix} x_1 \\ x_2 \\ x_3 \\ x_4 \\ x_5 \\ x_6 \\ x_7 \end{pmatrix} = \begin{pmatrix} x_3 + x_5 + x_7 \\ x_3 + x_6 + x_7 \\ x_3 \\ x_5 + x_6 + x_7 \\ x_5 \\ x_6 \\ x_7 \end{pmatrix} = x_3 \begin{pmatrix} 1 \\ 1 \\ 1 \\ 0 \\ 0 \\ 0 \\ 0 \end{pmatrix} + x_5 \begin{pmatrix} 1 \\ 0 \\ 0 \\ 1 \\ 1 \\ 0 \\ 0 \end{pmatrix} + x_6 \begin{pmatrix} 0 \\ 1 \\ 0 \\ 1 \\ 0 \\ 1 \\ 0 \end{pmatrix} + x_7 \begin{pmatrix} 1 \\ 1 \\ 0 \\ 1 \\ 0 \\ 0 \\ 1 \end{pmatrix},$$

which shows that every solution is a linear combination of $\{c_1, c_2, c_3, c_4\}$, i.e., these four vectors span the nullspace of H. The decomposition of the elements of C also demonstrates the linear independence of $\{c_1, c_2, c_3, c_4\}$ since, if a linear combination of these four vectors equals the zero vector, then the 3rd, 5th, 6th, and 7th coordinates of the result show that the four coefficients are all 0. This proves that $\{c_1, c_2, c_3, c_4\}$ forms a basis for C.

Example 4.21. Let V be a line through the origin in $\mathbb{R}^3$. Analytic geometry tells us that this line can be described as the set of points of the form $t(a, b, c)$, where t is an arbitrary scalar, and (a, b, c) is a fixed nonzero vector on the line. In particular, V is a subspace of $\mathbb{R}^3$ and $\{(a, b, c)\}$ forms a basis for V. First, from the analytic geometry description of V, it is obvious that $\{(a, b, c)\}$ spans. For independence, if $\alpha(a, b, c) = 0$, then since $(a, b, c) \neq \mathbf{0}$, one of the entries is nonzero. Since $(0, 0, 0) = (\alpha a, \alpha b, \alpha c)$, we have $\alpha a = \alpha b = \alpha c = 0$. Either $a = b = c = 0$, which is false, or $\alpha = 0$. More generally, every nonzero vector in a vector space constitutes a linearly independent set.

Example 4.22. Let V be a plane through the origin in $\mathbb{R}^3$. From multivariable calculus we know that the points (x, y, z) in V are those that satisfy the equation $ax + by + cz = 0$ for an appropriate choice of a, b, c not all equal to 0 and therefore V is a subspace of $\mathbb{R}^3$. Suppose we have fixed values of a, b, and c that determine V from this equation, and for convenience suppose that $a \neq 0$. Two specific elements of V are:

$$\begin{aligned} v &= (-b/a, 1, 0), \\ w &= (-c/a, 0, 1). \end{aligned}$$

These vectors, which were found by setting $y = 1$ and $z = 0$ and solving for x to find v, and setting $y = 0$ and $z = 1$, and solving for x to find w, constitute a basis for V. First, for independence, if α and β are scalars with $\alpha v + \beta w = \mathbf{0}$, then

$$\begin{aligned} (0, 0, 0) &= \alpha(-b/a, 1, 0) + \beta(-c/a, 0, 1) \\ &= (-\alpha b/a - \beta c/a, \alpha, \beta). \end{aligned}$$

Equating the second and third components yields $\alpha = 0$ and $\beta = 0$. Next, for spanning, let (x, y, z) be a point on V. Then $ax + by + cz = 0$. Solving for x, we have $x = (-b/a)y + (-c/a)z$. Thus,

$$\begin{aligned} (x, y, z) &= ((-b/a)y + (-c/a)z, y, z) = ((-b/a)y, y, 0) + ((-c/a)z, 0, z) \\ &= y(-b/a, 1, 0) + z(-c/a, 0, 1) = yv + zw, \end{aligned}$$

which says exactly that every point on V is a linear combination of v and w. Since $\{v, w\}$ spans V and is linearly independent, this set forms a basis of V.

Example 4.23. Let F be any field, and let $V = F^2$. Then an argument similar to that of the example of $\mathbb{R}^3$ above shows that $\{(1,0),(0,1)\}$ is a basis for V. However, we can produce lots of other bases for V. For example, let $v = (a,b)$ be a nonzero vector and $w = (c,d)$ any vector which is not a scalar multiple of v (this implies that $ad - bc \neq 0$). Then $\{v, w\}$ is a basis for V. From Exercise 8 below it suffices to show that $(1,0)$ and $(0,1)$ are in the span of $\{v, w\}$ or, in other words, that there are scalars x, y, z, w with

$$x(a,b) + y(c,d) = (1,0)$$
$$z(a,b) + w(c,d) = (0,1).$$

These equations can be written as two linear systems, the first of which is:

$$ax + yc = 1$$
$$bx + dy = 0.$$

The coefficient matrix for both systems is $A = \begin{pmatrix} a & c \\ b & d \end{pmatrix}$; assume that $a \neq 0$. (If $a = 0$, then $b \neq 0$ and switch the rows). Row operations reduce A to

$$\begin{pmatrix} 1 & c \\ 0 & d - \frac{cb}{a} \end{pmatrix} = \begin{pmatrix} 1 & c \\ 0 & \frac{ad-bc}{a} \end{pmatrix},$$

whence to $\begin{pmatrix} 1 & c \\ 0 & 1 \end{pmatrix}$. Clearly every linear system with coefficient matrix A has a unique solution, so that $(1,0)$ and $(0,1)$ are in the span of $\{v, w\}$.

From this example, we see that there are many bases for F^2. An argument analogous to the one just given shows that the columns of every $n \times n$ matrix of rank n constitute a basis for F^n and conversely. In particular, what is common to all the bases we constructed is that they contain the same number of elements. This is a special case of a more general fact which we state now. Recall that cardinality measures the size of a set. The cardinality of a finite set S is the number of elements in S. We refer the reader to any linear algebra text for a proof the following theorem.

Theorem 4.24. *Let F be a field and let V be an F-vector space. Then V has a basis. Moreover, the cardinalities of any two bases V are equal.*

Definition 4.25. Let F be a field and let V be an F-vector space with basis B. Then the dimension of V is the cardinality of B.

We will use the following result to determine the number of codewords of certain error correcting codes. The codes we will consider will be $\mathbb{Z}_2$-vector spaces.

Proposition 4.26. *Suppose that V is a $\mathbb{Z}_2$-vector space. If* $\dim(V) = n$, *then* $|V| = 2^n$.

Proof. We give two arguments for this. First, suppose that $\{v_1, \ldots, v_n\}$ is a basis for V. Then every element of V is of the form $a_1v_1 + \cdots + a_nv_n$ with scalars $a_i \in \mathbb{Z}_2$. So, to produce elements of V, we choose the scalars. Since $|\mathbb{Z}_2| = 2$, we have two choices for each a_i. Linear independence of $v_1, \ldots, v_n$ shows that distinct choices of the n scalars $a_1, \ldots, a_n$ yield distinct vectors so, by a standard counting principle, we have $2 \cdot 2 \cdot \cdots \cdot 2 = 2^n$ total choices for the a_i and hence$|V| = 2^n$.

For an alternative argument, recall that if X and Y are finite sets, then $|X \times Y| = |X| \cdot |Y|$. Using induction, it follows that $|\mathbb{Z}_2^n| = 2^n$. Now, given a basis $B = \{v_1, \ldots, v_n\}$ for V, the function $\mathbb{Z}_2^n \to V$ given by $(a_1, \ldots, a_n) \mapsto a_1 v_1 + \cdots + a_n v_n$ is onto since B spans V, and it is 1-1 since B is linearly independent. Consequently this function is a bijection and the two finite sets $\mathbb{Z}_2^n$ and V have the same size. We conclude that $|V| = 2^n$. □

A central feature of modern algebra is to investigate algebraic structures along with the appropriate mappings between them. For vector spaces these mappings are called linear transformations.

Definition 4.27. Let V and W be vector spaces over the scalar field F. A function $T : V \to W$ is called a linear transformation provided

1. $T(v_1 + v_2) = T(v_1) + T(v_2)$ for every pair of vectors v_1 and v_2 in V and
2. $T(av) = aT(v)$ for every scalar $a \in F$ and every $v \in V$.

The fundamental connection between linear transformations and matrices is effected by the notion of basis. Indeed, the selection of bases for the domain and range enables the realization of a linear transformation of finite dimensional vector spaces as a matrix. If $B_1 = \{v_1, \ldots, v_n\}$ is a basis for V and $B_2 = \{w_1, \ldots, w_m\}$ is a basis for W , then each $T(v_i)$ is uniquely expressed as the linear combination $T(v_j) = \sum_{j=1}^{m} a_{ij} w_i$. The linear transformation T is then identified with the $m \times n$ matrix (a_{ij}). This identification is basis dependent in the sense that the same linear transformation can have different matrix representations for different basis of V and W. To emphasize this fact, the unique matrix representation for the linear transformation T with respect to given bases B_1 and B_2 is denoted ${}_{B_1}[T]_{B_2}$.

In the special case that $V = W$ is finite dimensional and we have a fixed basis B for V, we write $[T]_B$ instead of ${}_B[T]_B$. We can then ask the following question: If B' is another basis for V, how are $[T]_B$ and $[T]_{B'}$ related? As noted above, if the elements of B' are written as linear combinations of the elements of B, then the coefficients of the vectors in B are the columns of an $n \times n$ matrix S of rank n. The matrix S is then invertible and we have the change of basis formula $[T]_{B'} = S[T]_{B'}S^{-1}$. The matrices $[T]_B$ and $[T]_{B'}$ are said to be similar. Two important scalars associated with square matrices are the determinant and the trace (the sum of the diagonal entries). We refer to a text on linear algebra for the basic properties of these quantities, but point out an important fact about them with respect to similar matrices:

Theorem 4.28. *Let A and B be similar $n \times n$ matrices. Then the traces of A and B are equal and so are their determinants.*

Exercises

1. Suppose that W is an m dimensional subspace of the finite dimensional vector space V over the field F. Explain why the following procedure will produce a basis for W. Recall that the span of a collection of vectors in a vector space is the subspace consisting of all of their linear combinations: $\text{Span}\{w_1, \ldots, w_k\} = \{\sum_{i=1}^{k} \alpha_i w_i : \alpha_i \in F\}$.
   (a) Find a nonzero vector w_1 in W and construct $\text{Span}\{w_1\}$.
   (b) For each $i = 2, \ldots, m-1$ find a vector $w_{i+1} \in W - \text{Span}\{w_1, \ldots, w_i\}$.
   (c) Then $\{w_1, \ldots, w_m\}$ is a basis for W.
2. If $w = (1, 2, 3)$, find a basis for the subspace W of Exercise 1 of Sect. 4.1.

3. Referring to Sect. 2.2, explain why the following procedure is an algorithm to find a basis for the span W of a given set of m vectors in F^n, where F is a field:

   (a) Form the matrix A whose rows are the given vectors.
   (b) Reduce A to row reduced echelon form matrix B.
   (c) The nonzero rows of B constitute a basis for W.

4. Find a basis for the $\mathbb{Z}_2$-vector space $C = \{00000, 11010, 01101, 10111\}$.
5. Find a basis for the $\mathbb{Z}_2$-vector space

$$C = \{000000, 110011, 101110, 101001, 110100, 011010, 011101, 000111\}$$

   and express each element of C as a linear combination of the basis vectors.
6. Let C be the $\mathbb{Z}_2$- subspace of $\mathbb{Z}_2^5$ spanned by the vectors 11110, 10101, 01011, 11011, 10000. Find a basis for C.
7. Let W be a subspace of the finite dimensional vector space V. If $\dim(W) = \dim(V)$, show that $W = V$. Conclude that if W is a proper subspace of V (meaning that W is a proper subset of V), then $\dim(W) < \dim(V)$.
8. Let F be a field. Construct a proper subspace W of $F[x]$ for which $\dim(W) = \dim(F[x])$.
9. Let V be a vector space over $\mathbb{Z}_2$.

   (a) If $u, v \in V$ are both nonzero and distinct, show that $\{u, v\}$ is linearly independent.
   (b) If $u, v, w \in V$ are all nonzero and distinct, show that $\{u, v, w\}$ is linearly independent if $w \neq u + v$.

10. Let $\{v_1, \dots, v_k\}$ be a set of linearly independent vectors in an F-vector space V and let $w \in V$. If w is not in the span of $\{v_1, \dots, v_k\}$, show that $\{v_1, \dots, v_k, w\}$ is linearly independent.
11. Suppose that $\{v_1, \dots, v_k\}$ is a basis for the vector space V and $w_1, \dots, w_k \in V$ have the property that each of the v_i is in the span of $\{w_1, \dots, w_k\}$. Prove that $\{w_1, \dots, w_k\}$ is a basis for V.
12. Let C be the nullspace in $\mathbb{Z}_2^8$ of the matrix

$$\begin{pmatrix} 0\,0\,0\,1\,1\,1\,1\,0 \\ 0\,1\,1\,0\,0\,1\,1\,0 \\ 1\,0\,1\,0\,1\,0\,1\,0 \\ 1\,1\,1\,1\,1\,1\,1\,1 \end{pmatrix}.$$

   Determine the codewords in C, its dimension, distance, and error correction capability.

4.3 Linear Codes

A *linear code* is a code C which, as a subset of $\mathbb{Z}_2^n$, is subspace of $\mathbb{Z}_2^n$.

Example 4.29. If A is an $m \times n$ matrix over $\mathbb{Z}_2$, then the nullspace of A is a linear code of length n, since we know that the nullspace of a matrix is always a vector space. In particular, the Hamming code and the Golay code are linear codes.

Example 4.30. The subset $\{(1,0), (0,1)\}$ of $\mathbb{Z}_2^2$ is not a linear code. The smallest linear code containing $\{(1,0), (0,1)\}$ is the span of this set, which is equal to $\mathbb{Z}_2^2$. Similarly, $\{11100, 00111\}$ is not a linear code. Its span is a linear code, and the span is $\{00000, 11100, 00111, 11011\}$.

Recall that the distance d of a code is defined to be the minimum distance between distinct codewords:

$$d = \min\{D(u,v) : u, v \in C, u \neq v\},$$

and that the distance $D(u,v)$ between any pair (u,v) of words is calculated as $D(u,v) = \mathrm{wt}(u+v)$.

Lemma 4.31. *If C is a linear code with distance d, then*

$$d = \min\{\mathrm{wt}(v) : v \in C, v \neq \mathbf{0}\}.$$

Proof. Let $e = \min\{\mathrm{wt}(v) : v \in C, v \neq \mathbf{0}\}$. We may write $e = \mathrm{wt}(u)$ for some nonzero vector $u \in C$. Then $e = D(u, \mathbf{0})$. Since $\mathbf{0} \in C$, we see that $d \leq e$ by definition of d. Conversely, we may write $d = D(v,w)$ for some distinct $v, w \in C$. Then $d = \mathrm{wt}(v+w)$. Since C is a linear code, $v + w \in C$, and $v + w \neq \mathbf{0}$ since $v \neq w$. Therefore, by definition of e, we see that $e \leq d$. Thus, $d = e$, as desired. □

This lemma simplifies finding the distance of a linear code. For example, the Hamming code is

$$C = \{1111111, 0011001, 0000000, 1110000, 1101001, 1100110, 0110011, 0100101,$$
$$0010110, 0111100, 1011010, 1010101, 1000011, 0101010, 1001100, 0001111\}.$$

A quick inspection shows that the smallest weight of a nonzero codeword of C is 3; therefore, $d = 3$.

Let C be a linear code of length n and distance d. If the dimension of C is k, then we refer to C as an (n,k,d) linear code. Any matrix whose rows constitute a basis for C is called a *generator matrix* for C. Thus a generator matrix for an (n,k,d) linear code is an $n \times k$ matrix whose rank is k. The rows of any such matrix G span an (n,k,d) linear code C, and any matrix row equivalent to G is also a generator matrix for C.

The reason for the terminology "generator matrix" is the following theorem.

Theorem 4.32. *Let C be an (n,k,d) linear code with generator matrix G. Then $C = \{vG : v \in \mathbb{Z}_2^k\}$.*

Proof. For $v \in \mathbb{Z}_2^k$, thought of as a row vector, vG is a linear combination of the rows of G. Allowing v range over all of $\mathbb{Z}_2^k$ results in the span of the rows of G which is, by definition, C. □

For any row vector $v \in \mathbb{Z}_2^k$, the vector vG is the linear combination of the rows of G whose coefficients are the entries of v. Therefore $\{vG : v \in \mathbb{Z}_2^k\}$ is the row space of G which, by definition, is C.

The generator matrix thus "encodes" v as vG in C. The $n - k$ "extra" coordinates in vG are analogous to the check digit in the identification numbers discussed in Chap. 1.

Example 4.33. Recall that the Hamming code C is defined to be the nullspace of the matrix

$$H = \begin{pmatrix} 0 & 0 & 0 & 1 & 1 & 1 & 1 \\ 0 & 1 & 1 & 0 & 0 & 1 & 1 \\ 1 & 0 & 1 & 0 & 1 & 0 & 1 \end{pmatrix}.$$

A matrix calculation shows that the solutions to the system equation $HX = \mathbf{0}$, written as columns, are the vectors of the form

$$\begin{pmatrix} x_1 \\ x_2 \\ x_3 \\ x_4 \\ x_5 \\ x_6 \\ x_7 \end{pmatrix} = x_3 \begin{pmatrix} 1 \\ 1 \\ 1 \\ 0 \\ 0 \\ 0 \\ 0 \end{pmatrix} + x_5 \begin{pmatrix} 1 \\ 0 \\ 0 \\ 1 \\ 1 \\ 0 \\ 0 \end{pmatrix} + x_6 \begin{pmatrix} 0 \\ 1 \\ 0 \\ 1 \\ 0 \\ 1 \\ 0 \end{pmatrix} + x_7 \begin{pmatrix} 1 \\ 1 \\ 0 \\ 1 \\ 0 \\ 0 \\ 1 \end{pmatrix}.$$

The four vectors in the right hand side of this equation, written as rows, form a basis for C. That is,

$$\{1110000, 1001100, 0101010, 1101001\}$$

is a basis for C. Thus, the matrix

$$\begin{pmatrix} 1\,1\,1\,0\,0\,0\,0 \\ 1\,0\,0\,1\,1\,0\,0 \\ 0\,1\,0\,1\,0\,1\,0 \\ 1\,1\,0\,1\,0\,0\,1 \end{pmatrix}$$

is a generator matrix for C. Alternatively, if we row reduce this matrix, we obtain another generator matrix. Doing so, the matrix

$$G = \begin{pmatrix} 1\,0\,0\,0\,0\,1\,1 \\ 0\,1\,0\,0\,1\,0\,1 \\ 0\,0\,1\,0\,1\,1\,0 \\ 0\,0\,0\,1\,1\,1\,1 \end{pmatrix}$$

is another generator matrix. Thus, we may describe the Hamming code as $C = \{vG : v \in \mathbb{Z}_2^4\}$. We may view encoding a four-tuple v as vG as appending three check digits to v, since for any v, the product vG has v as the first 4 components; this is because the left 4×4 submatrix of G is the identity matrix. For example, if $v = 1011$, then $vG = 1011010$.

Thus, by using a generator matrix in row reduced echelon form, we may view encoding v as vG as adding a certain number (which is $n-k$) of extra digits to v. The "redundant" extra digits introduced in the construction of linear codes generalize the "redundant" check digit that enables the error detection in identification number schemes discussed in the first chapter.

We defined a generator matrix of an (n, k, d) linear code C to be a matrix G whose rows form a basis for C. Thus G is a $k \times n$ matrix of rank k. Reversing this line of thought, we can instead produce linear codes from matrices.

Remark 4.34. While it is standard to use the notation $a_1 \cdots a_n$ for words in a code of length n (e.g. 1011010) it is unnatural and perhaps confusing to use this notation for vectors in a subspace of the vector space F^m in theoretical discussions. The remainder of this section uses the standard linear algebra notation of "tuples" for vectors.

Proposition 4.35. *Let G be a $k \times n$ matrix over $\mathbb{Z}_2$ of rank k. If $C = \{vG : v \in \mathbb{Z}_2^k\}$, then C is an (n, k, d) linear code for some d and G is a generator matrix for C.*

Proof. It follows from matrix multiplication that if $v = (a_1, \ldots, a_k) \in \mathbb{Z}_2^k$ and $r_1, \ldots, r_k$ are the k rows of G, then $vG = \sum_i a_i r_i$. Moreover, this equation shows that any linear combination of rows of G can be written in the form vG. Thus C is the span of the rows of C so that C is a subspace of $\mathbb{Z}_2^n$. Moreover, since the rows of G are linearly independent, they form a basis for C. Thus C has dimension k and, if d is the distance of C, then C is an (n, k, d) code. By construction, G is a generator matrix for C. □

We initially constructed codes as nullspaces of matrices. We would like to make some simple restrictions on the matrices we use to build codes in this way. If C is an (n, k, d) code, then a matrix H is called a *parity check matrix* for C if H is an $(n-k) \times n$ matrix whose nullspace is exactly C, with its elements written as columns. Alternatively, we may say that H is a parity check matrix for C if C is the nullspace of H and if the rows of H are linearly independent. We leave the proof of the equivalence of these two statements as an exercise. To keep straight the difference between vectors and column matrices, if H is a parity check matrix for C, then by writing A^T for the transpose of a matrix,

$$C = \left\{v \in \mathbb{Z}_2^n : Hv^T = \mathbf{0}\right\}.$$

Example 4.36. The Hamming matrix H is a parity check matrix for the Hamming code C, since C is the nullspace of H and since the rows of H are seen to be linearly independent. Similarly, the matrix $A = [I : B]$ defined in Sect. 2.5 is a parity check matrix for the Golay code because the code is the nullspace of A and the 12 rows of A are linearly independent.

The generator and parity check matrices of a code both have the property that their rows are linearly independent. Thus, a parity check matrix of a code can be taken to be a generator matrix of another code. We can be more specific about this other code. To do so, we define an analogue of the dot product. For $u = (u_1, \ldots, u_n), v = (v_1 \ldots, v_n) \in \mathbb{Z}_2^n$, define $u \cdot v$ by

$$u \cdot v = u_1 v_1 + \cdots + u_n v_n.$$

It is an easy exercise to verify that if $u, v, w \in \mathbb{Z}_2^n$ and $a \in \mathbb{Z}_2$, then $(u + w) \cdot v = u \cdot v + w \cdot v$, and $u \cdot (av) = a(u \cdot v)$.

If C is a code, we define the dual code

$$C^\perp = \{v \in \mathbb{Z}_2^n : v \cdot c = 0 \text{ for all } c \in C\}.$$

Lemma 4.37. *Let C be a linear code. Then $C^\perp$ is a linear code. Moreover, if $\{c_1, \ldots, c_k\}$ is a basis for C, then $v \in C^\perp$ if and only if $v \cdot c_i = 0$ for each i.*

Proof. To show that $C^\perp$ is linear we must show that it is closed under addition. Let $v, w \in C^\perp$. Then, for each $c \in C$, we have $v \cdot c = 0$ and $w \cdot c = 0$. Therefore, $(v+w)\cdot c = v\cdot c + w \cdot c = 0+0 = 0$. Thus, $v + w \in C^\perp$. Next, let $\{c_1, \ldots, c_k\}$ be a basis for C. If $v \in C^\perp$, then $v \cdot c = 0$ for all $c \in C$. Since the c_i are elements of C, this implies that $v \cdot c_i = 0$ for all i. Conversely, if v is a word such that $v \cdot c_i = 0$ for all i, then let $c \in C$. We may write $c = a_1 c_1 + \cdots + a_k c_k$ for some $a_i \in \mathbb{Z}_2$. Then

$$\begin{aligned} v \cdot c &= v \cdot (a_1 c_1 + \cdots + a_k c_k) = v \cdot (a_1 c_1) + \cdots + v \cdot (a_k c_k) \\ &= a_1 (v \cdot c_1) + \cdots + a_k (v \cdot c_k) = a_1 0 + \cdots + a_k 0 = 0. \end{aligned}$$

Thus, $v \in C^\perp$. This proves that $C^\perp = \left\{v \in \mathbb{Z}_2^n : v \cdot c_i = 0 \text{ for each } i\right\}$. □

Theorem 4.38. *Let C be a linear code of length n and dimension k with generator matrix G and parity check matrix H.*

1. $HG^T = 0$.
2. *H is a generator matrix for $C^\perp$ and G is a parity check matrix for $C^\perp$.*
3. *The linear code $C^\perp$ has dimension $n - k$.*
4. $(C^\perp)^\perp = C$.
5. *If H' is a generator matrix for $C^\perp$ and G' is a parity check matrix for $C^\perp$, then H' is a parity check matrix for C and G' is a generator matrix for C.*

Proof. Let $v \in \mathbb{Z}_2^k$. Then $vG \in C$. Thus, by definition of parity check matrix, $H(vG)^T = \mathbf{0}$. But, $(vG)^T = G^T v^T$. Therefore, $(HG^T)v^T = \mathbf{0}$. Since v can be any element of $\mathbb{Z}_2^k$, we see that $(HG^T)x = \mathbf{0}$ for any column vector x. If e_i is the column vector with a 1 in the ith coordinate and 0 elsewhere, we see that $\mathbf{0} = (HG^T)e_i$, and this product is the ith column of HG^T. Since this is true for all i, we see that $HG^T = 0$. This proves the first statement.

For the second, we note that we only need to prove that $C^\perp$ is the nullspace of G and that the rows of H form a basis of $C^\perp$. For the first part, let $v \in C^\perp$. To see that v is in the nullspace, we must see that $Gv^T = \mathbf{0}$. However, the ith entry of Gv^T is the product of the ith row r_i of G with v^T. In other words, it is $r_i \cdot v$. However, since the rows of G form a basis for C, we see that $r_i \cdot v = 0$. Thus, $Gv^T = \mathbf{0}$. Conversely, if v is in the nullspace of G, then $Gv^T = \mathbf{0}$. This means that $r_i \cdot v = 0$ for all i. The rows of G form a basis for C; thus, by the lemma, $v \in C^\perp$.

As a consequence of knowing that the nullspace of G is $C^\perp$, we prove Statement 3. From the rank plus nullity theorem, n is the sum of the dimensions of the nullspace and the row space of G. By definition, the row space of G is C, so its dimension is k. Thus, the nullspace has dimension $n - k$. But, this space is $C^\perp$. So, $\dim(C^\perp) = n - k$. To finish the proof of Statement 2 we show that the rows of H form a basis of $C^\perp$. Because we know that the rows are independent, we only need to see that they span $C^\perp$. We first show that the rows are elements of $C^\perp$. Since H is a parity check matrix for C, if $v \in C$, then $Hv^T = \mathbf{0}$. The ith entry of Hv^T is the dot product of the ith row of H with v; thus, this dot product is 0. Since this is true for all $v \in C$, we see that the rows of H are all elements of $C^\perp$. So, the row space of $H^\perp$ is contained in $C^\perp$. However, both of these spaces have dimension $n - k$; we are using the assumption that the rows of H are linearly independent to see that the row space of H has dimension $n - k$. It is an exercise to show that if a subspace of a vector space has the same dimension as the vector space, then they are equal. From this we conclude that $C^\perp$ is the row space of H. Thus, the rows of H form a basis for $C^\perp$, which says that H is a generator matrix for $C^\perp$.

To prove (4), we note that the inclusion $C \subseteq (C^\perp)^\perp$ follows from the definitions: If $c \in C$, then $c \cdot v = 0$ for all $v \in C^\perp$ by definition of $C^\perp$. This is what we need to see that $c \in (C^\perp)^\perp$ However, from (3) we see that the dimension of $(C^\perp)^\perp$ is $n - (n - k) = k$, which is the dimension of C. From the inclusion $C \subseteq (C^\perp)^\perp$ we then conclude $C = (C^\perp)^\perp$. Finally, (5) follows from the previous statements, since if H' is a generator matrix for $C^\perp$, then it is a parity check matrix for $(C^\perp)^\perp = C$, and if G' is a parity check matrix for $C^\perp$, then it is a generator matrix for $(C^\perp)^\perp = C$. □

Exercises

1. If $C \subseteq \mathbb{Z}_2^n$ is a code with basis $\{c_1, \ldots, c_k\}$ show that

$$C^\perp = \{v \in \mathbb{Z}_2^n : v \cdot c_i = 0 \text{ for all } i\}.$$

2. Construct a nontrivial code C for which $C = C^\perp$.
3. Find a basis for the Hamming code C, and use it to determine the dual code $C^\perp$. What is its distance and error correcting capability?

4. Find a generator matrix G and a parity check matrix H for the code $C^{\perp}$ of the previous problem. Check that $HG^T = 0$. Then prove that a parity check matrix and generator matrix for any code must satisfy this property.
5. Let

$$G = \begin{pmatrix} 1\,1\,1\,1\,0\,0\,0\,0 \\ 0\,0\,0\,0\,1\,1\,1\,1 \\ 1\,0\,1\,0\,1\,0\,1\,0 \end{pmatrix}$$

be a generator matrix for a code C. Write out the codewords of C and find a parity check matrix for C.

References

1. Halmos P (1987) Finite dimensional vector spaces, 2nd edn. Undergraduate texts in mathematics. Springer, New York
2. Herstein I (1975) Topics in algebra, 2nd edn. Wiley, Hoboken

Chapter 5
Quotient Rings and Field Extensions

In this chapter we describe a method for producing new rings from a given one. Of particular interest for applications is the case of a field extension of a given field. If F is a field, then a *field extension* is a field K that contains F and for which the field operations restrict to the field operation of F. In this context F is also referred to as a *subfield* of K. For example, $\mathbb{C}$ is a field extension of $\mathbb{R}$ since $\mathbb{C}$ is a field containing $\mathbb{R}$ and the field operations on the real elements of $\mathbb{C}$ are precisely those of the field $\mathbb{R}$. Similarly, $\mathbb{C}$ and $\mathbb{R}$ are field extensions of $\mathbb{Q}$, and we can view $\mathbb{R}$ and $\mathbb{Q}$ as subfields of $\mathbb{C}$. The term *intermediate field extension* is used for $\mathbb{R}$ in this situation since $\mathbb{Q} \subset \mathbb{R} \subset \mathbb{C}$. When we consider ruler and compass constructions, we will need to investigate intermediate field extensions of the extension $\mathbb{Q} \subset \mathbb{R}$. For coding theory we need field extensions of $\mathbb{Z}_2$.

The extensions of a field F that we need are called finite algebraic field extensions and they are produced in an analogous fashion to the construction of the integers modulo a fixed integer n. Here the ring $\mathbb{Z}$ is replaced by the ring $F[x]$ of polynomials in the indeterminate x and the integer n by a polynomial $f(x) \in F[x]$. In order to do this we need to know that the arithmetic of polynomials is sufficiently similar to the arithmetic of integers. In the first section of this chapter we see that notions relating to divisibility work just as well for polynomials over a field as for the integers. Next the construction of the integers modulo n is generalized to arbitrary commutative rings. Finally these ideas come together in the construction of algebraic field extensions.

5.1 Arithmetic of Polynomial Rings

Let F be a field, and let $F[x]$ be the ring of polynomials in the indeterminate x. High school students study the arithmetic of this ring without saying so in so many words, at least for the case $F = \mathbb{R}$. In this section we make a formal study of this arithmetic, seeing that much of what we did for integers above can be done in the ring $F[x]$. We start with the most basic definition.

Definition 5.1. Let f and g be polynomials in $F[x]$. Then we say that f divides g, or g is divisible by f, if there is a polynomial $h \in F[x]$ with $g = fh$.

Electronic supplementary material The online version of this chapter (doi:10.1007/978-3-319-04498-9_5) contains supplementary material, which is available to authorized users. The supplementary material can also be downloaded from http://extras.springer.com.

D.R. Finston and P.J. Morandi, *Abstract Algebra: Structure and Application*,
Springer Undergraduate Texts in Mathematics and Technology, DOI 10.1007/978-3-319-04498-9_5

The greatest common divisor of two integers a and b is the largest integer dividing both a and b. This definition needs to be modified a little for polynomials. While we cannot talk about "largest" polynomial in the same manner as for integers, we can talk about the degree of a polynomial. Recall that the *degree* of a nonzero polynomial f is the largest integer m for which the coefficient of x^m is nonzero. If $f(x) = a_n x^n + a_{n-1}x^{n-1} + \cdots + a_1 x + a_0$ and $a_n \neq 0$, then the degree of $f(x)$ is n. We write $\deg(f)$ for the degree of f. The degree function allows us to measure the size of polynomials. However, there is one extra complication. For example, any polynomial of the form ax^2 with $a \neq 0$ divides x^2 and x^3. Thus, there isn't a unique polynomial of highest degree that divides a pair of polynomials. To pick one out, we consider *monic* polynomials, whose *leading coefficient* is 1. For example, x^2 is the monic polynomial of degree 2 that divides both x^2 and x^3, while $5x^2$ is not monic. As a piece of terminology, we will refer to an element $f \in F[x]$ as a *polynomial over* F.

Definition 5.2. Let f and g be polynomials over F, not both zero. Then a greatest common divisor of f and g is a monic polynomial of largest degree that divides both f and g.

The problem with the definition above has to do with uniqueness. Could there be more than one greatest common divisor of a pair of polynomials? The answer is no, and we will prove this after we prove the analogue of the Division Algorithm. In fact the reason for working with polynomials having coefficients in a field rather than a more general ring is to ensure that the Division Algorithm is valid. Before we state and prove it, we need a simple lemma about degrees. For convenience, we set $\deg(0) = -\infty$. We also make the convention that $-\infty + -\infty = -\infty$ and $-\infty + n = -\infty$ for any integer n. The point of these conventions is to make the statement in the following lemma and other results as simple as possible.

Lemma 5.3. *Let F be a field and let f and g be polynomials over F. Then* $\deg(fg) = \deg(f) + \deg(g)$.

Proof. If either $f = 0$ or $g = 0$, then the equality $\deg(fg) = \deg(f) + \deg(g)$ is true by our convention above. So, suppose that $f \neq 0$ and $g \neq 0$. Write $f = a_n x^n + \cdots + a_0$ and $g = b_m x^m + \cdots + b_0$ with $a_n \neq 0$ and $b_m \neq 0$. Therefore, $\deg(f) = n$ and $\deg(g) = m$. The definition of polynomial multiplication yields

$$fg = (a_n b_m)x^{n+m} + (a_n b_{m-1} + a_{n-1} b_m)x^{n+m-1} + \cdots + a_0 b_0.$$

Now, since the coefficients come from a field, which has no zero divisors, we can conclude that $a_n b_m \neq 0$, so $\deg(fg) = n + m = \deg(f) + \deg(g)$, as desired. □

Proposition 5.4 (Division Algorithm). *Let F be a field and let f and g be polynomials over F with f nonzero. Then there are unique polynomials q and r with $g = qf + r$ and* $\deg(r) < \deg(f)$.

Proof. The existence of q and r is clear if $g = 0$ since we can set $q = r = 0$. So, assume that $g \neq 0$. Let

$$S = \{t \in F[x] : t = g - qf \text{ for some } q \in F[x]\};$$

this is a nonempty set of nonzero polynomials, since $g \in S$. By the well-ordering property of the integers, there is a polynomial r of least degree in S and the defining property of S guarantees a $q \in F[x]$ with $r = g - qf$, so $g = qf + r$. We need to see that $\deg(r) < \deg(f)$. If, on the other hand, $\deg(r) \geq \deg(f)$, say $n = \deg(f)$ and $m = \deg(r)$. If $f = a_n x^n + \cdots + a_0$ and $r = r_m x^m + \cdots + r_0$ with $a_n \neq 0$ and $r_m \neq 0$, then by thinking about the method of long division of polynomials, we realize that we may write $r = (r_m a_n^{-1})x^{m-n} f + r'$ with $\deg(r') < m = \deg(r)$. But then

$$g = qf + r = qf + (r_m a_n^{-1})x^{m-n} f + r' = (q + r_m a_n^{-1} x^{m-n})f + r',$$

which shows that $r' \in \mathcal{S}$. Since $\deg(r') < \deg(r)$, this would be a contradiction to the choice of r. Therefore, $\deg(r) \geq \deg(f)$ is false, so $\deg(r) < \deg(f)$, as we wanted to prove. This proves existence of q and r. For uniqueness, suppose that $g = qf + r$ and $g = q'f + r'$ for some polynomials q, q' and r, r' in $F[x]$, and with $\deg(r), \deg(r') < \deg(f)$. Then $qf + r = q'f + r'$, so $(q - q')f = r' - r$. Taking degrees and using the lemma, we have

$$\deg(q' - q) + \deg(f) = \deg(r' - r).$$

Since $\deg(r) < \deg(f)$ and $\deg(r') < \deg(f)$, we have $\deg(r' - r) < \deg(f)$. However, if $\deg(q' - q) \geq 0$, this is a contradiction to the equation above. The only way for this to hold is for $\deg(q' - q) = \deg(r' - r) = -\infty$. Thus, $q' - q = 0 = r' - r$, so $q' = q$ and $r' = r$, proving uniqueness. □

We now prove the existence of greatest common divisors of polynomials, and also prove the representation theorem analogous to Proposition 1.13.

Proposition 5.5. *Let F be a field and let f and g be polynomials over F, not both zero. Then* $\gcd(f, g)$ *exists and is unique. Furthermore, there are polynomials h and k with* $\gcd(f, g) = hf + kg$.

Proof. We will prove this by proving the representation result. Let

$$\mathcal{S} = \{hf + kg : h, k \in F[x]\}.$$

Then $\mathcal{S}$ contains nonzero polynomials as $f = 1 \cdot f + 0 \cdot g$ and $g = 0 \cdot f + 1 \cdot g$ both lie in $\mathcal{S}$. Therefore, the well-ordering property guarantees the existence of a nonzero polynomial $d \in \mathcal{S}$ of smallest degree. Write $d = hf + kg$ for some $h, k \in F[x]$. By dividing h and k by the leading coefficient of d, we may assume that d is monic without changing the condition $d \in \mathcal{S}$. We claim that $d = \gcd(f, g)$. To show that d is a common divisor of f and g, first consider f. By the Division Algorithm, we may write $f = qd + r$ for some polynomials q and r, and with $\deg(r) < \deg(d)$. Then

$$\begin{aligned} r &= f - qd = f - q(hf + kg) \\ &= (1 - qh)f + (-qk)g. \end{aligned}$$

This shows $r \in \mathcal{S}$. If $r \neq 0$, this would be a contradiction to the choice of d since $\deg(r) < \deg(d)$. Therefore, $r = 0$, which shows that $f = qd$, and so d divides f. Similarly, d divides g. Thus, d is a common divisor of f and g. If e is any other common divisor of f and g, then e divides any combination of f and g; in particular, e divides $hf + kg = d$. This forces $\deg(e) \leq \deg(d)$ by Lemma 5.3. Thus, d is the monic polynomial of largest degree that divides f and g, so d is a greatest common divisor of f and g. This proves everything but uniqueness. For that, suppose that d and d' are both monic common divisors of f and g of largest degree. By the proof above, we may write both d and d' as combinations of f and g. Applying the Division Algorithm to d and d' as above shows that d divides d' and vice-versa. If $d' = ad$ and $d = bd'$, then $d = bd' = abd$. Taking degrees shows that $\deg(ab) = 0$, which means that a and b are both constants. But, since d and d' are monic, for $d' = ad$ to be monic, $a = 1$. Thus, $d' = ad = d$. □

Exercises

1. Let F be a field and let f, g, q, r be polynomials in $F[x]$ such that $g = qf + r$. Prove that $\gcd(f, g) = \gcd(f, r)$.
2. Let F be a field. If $f \in F[x]$ has a multiplicative inverse in $F[x]$, prove that $\deg(f) = 0$. Conversely, show that any nonzero polynomial of degree 0 in $F[x]$ has a multiplicative inverse in $F[x]$.
   (This problem shows that the units of $F[x]$ are exactly the nonzero constant polynomials.)
3. Calculate, by using the Euclidean Algorithm, the greatest common divisor in $\mathbb{R}[x]$ of $x^5 + 5x^3 + x^2 + 4x + 1$ and $x^4 - x^3 - x - 1$, and write the greatest common divisor as a linear combination of the two polynomials. You may check your work with Maple, but do the calculation by hand.
4. Calculate and express the greatest common divisor of $x^{15} + x^{14} + x^{13} + x^{12} + x^{11} + x^{10} + x^9 + x^8 + x^7 + x^6 + x^5 + x^4 + x^3 + x^2 + x + 1$ and $x^9 - x^6 - x^4 - x^2 - x - 1$ as a linear combination of the two polynomials. You are welcome and encouraged to do this in Maple; if you do so, look at the worksheet Section-5.1-Exercise-4.mw.
5. Prove the *Remainder Theorem*, which asserts the following: Let $f(x)$ be a polynomial over F and $a \in F$. If $f(x) = q(x)(x - a) + r(x)$ according to the Division Algorithm, then $r(x)$ is equal to the constant polynomial $f(a)$. Conclude that a is a root of f if and only if $x - a$ divides $f(x)$.
   (Hint: Use the following results: if g and h are polynomials over F and $a \in F$, then $(gh)(a) = g(a)h(a)$ and $(g + h)(a) = g(a) + h(a)$.)
6. Let $f(x) = (x - a)g(x)$ for some $a \in F$ and some $g \in F[x]$. If $b \neq a$ is a root of f, show that $g(b) = 0$.
7. Prove that a polynomial of degree n has at most n roots in F.
   (Hint: Use induction together with the previous problems.)
8. Let $f \in F[x]$ be a polynomial of degree 2 or 3. Prove that f can be factored as $f = gh$ nontrivially (i.e., the degrees of both f and g are positive) if and only if f has a root in F.
9. Give an example of a polynomial $f \in \mathbb{R}[x]$ of degree 4 that can be factored nontrivially as $f = gh$ but for which f does not have a root in $\mathbb{R}$.
10. *Rational Roots Test*: Let $f(x) = a_n x^n + \cdots + a_0$ be a polynomial with integer entries. If $r = b/c$ is a rational root of f written in reduced form, show that b divides a_0 and c divides a_n.
    (Hint: Clear denominators in the equation $f(b/c) = 0$.)
11. Factor completely and determine the roots of the polynomial $x^5 - 8x^4 - 4x^3 + 76x^2 - 5x + 84$ as

    (a) a polynomial over $\mathbb{R}$;
    (b) a polynomial over $\mathbb{C}$;
    (c) a polynomial over $\mathbb{Z}_2$.

    (Feel free to do this with Maple; the file Section-5.1-Exercise-11.mw has some information about factoring polynomials over different fields.)

5.2 Ideals and Quotient Rings

We will construct extension fields of a field F in analogy with the construction of the rings $\mathbb{Z}_n$. More generally we will study the relevant structure in general commutative rings R and then apply it to the ring $F[x]$. Recall that $\mathbb{Z}_n$ consists of equivalence classes under the relation $a \sim b$ iff $b - a \in \{nt : t \in \mathbb{Z}\} = n\mathbb{Z}$. Replacing $\mathbb{Z}$ by R, the structure analogous to $n\mathbb{Z}$ is called an ideal of R, which we proceed to define.

Definition 5.6. Let R be a ring. An ideal I is a nonempty subset of R such that (i) if $a, b \in I$, then $a + b \in I$, and (ii) if $a \in I$ and $r \in R$, then $ar \in I$ and $ra \in I$.

This definition says that an ideal is a subset of R closed under addition that satisfies a strengthened form of closure under multiplication. Not only is the product of two elements of I also in I, but that the product of an element of I and any element of R is an element of I.

Example 5.7. For $R = \mathbb{Z}$ and n an integer, the set $n\mathbb{Z}$ is an ideal. For closure under addition, recall that $x, y \in n\mathbb{Z}$ means that there are integers a and b with $x = na$ and $y = nb$. Therefore $x+y = na+nb = n(a+b)$ so that $x+y \in n\mathbb{Z}$. For the multiplicative property, let $x = na \in n\mathbb{Z}$ and let $r \in \mathbb{Z}$. Then $rx = xr = r(na) = n(ra) \in n\mathbb{Z}$. This proves that $n\mathbb{Z}$ is an ideal. If $n > 0$, notice that

$$n\mathbb{Z} = \{0, n, 2n, \ldots, -n, -2n, \ldots\}$$

is the same as the equivalence class of 0 under the relation congruence modulo n. This is an important connection that we will revisit.

Example 5.8. Let $R = F[x]$ be the ring of polynomials over a field, and let $f \in F[x]$. Let

$$I = \{gf : g \in F[x]\},$$

the set of all multiples of f. This set is an ideal of $F[x]$. For closure under addition, let $h, k \in I$. Then $h = gf$ and $k = g'f$ for some polynomials g and g'. Then $h+k = gf+g'f = (g+g')f$, a multiple of f, so $h + k \in I$. For multiplication, let $h = gf \in I$, and let $a \in F[x]$. Then $ah = ha = agf = (ag)f$, a multiple of f, so $ah \in I$. Thus I is an ideal of $F[x]$.

Example 5.9. Let R be any commutative ring, and let $a \in R$. Let

$$aR = \{ar : r \in R\},$$

the set of all multiples of a. We show that aR is an ideal of R. First, let $x, y \in aR$. Then $x = ar$ and $y = as$ for some $r, s \in R$. Then $x + y = ar + as = a(r + s)$, so $x + y \in aR$. Next, let $x = ar \in aR$ and let $z \in R$. Then $xz = arz = a(rz) \in aR$. Also, $zx = xz$ since R is commutative, so $zx \in aR$. Therefore, aR is an ideal of R. This construction generalizes the previous two examples. The ideal aR is typically called the ideal *generated by a*, or the principal ideal generated by a. It is often written as (a).

Example 5.10. Let R be any commutative ring, and let $a, b \in R$. Set

$$I = \{ar + bs : r, s \in R\}.$$

To see that I is an ideal of R, first let $x, y \in I$. Then $x = ar + bs$ and $y = ar' + bs'$ for some $r, s, r', s' \in R$. Then

$$\begin{aligned} x + y &= (ar + bs) + (ar' + bs') \\ &= (ar + ar') + (bs + bs') \\ &= a(r + r') + b(s + s') \in I \end{aligned}$$

by the associative and distributive properties. Next, let $x \in I$ and $z \in R$. Again, $x = ar + bs$ for some $r, s \in R$. Then

$$\begin{aligned} xz &= (ar + bs)z = (ar)z + (bs)z \\ &= a(rz) + b(sz). \end{aligned}$$

This calculation shows that $xz \in I$. Again, since R is commutative, $zx = xz$, so $zx \in I$. Thus, I is an ideal of R. We can generalize this example to any finite number of elements of R: given $a_1, \ldots, a_n \in R$, if

$$J = \{a_1 r_1 + \cdots + a_n r_n : r_i \in R \text{ for each } i\},$$

then a similar argument will show that J is an ideal of R. The ideal J is typically referred to as the ideal generated by the elements $a_1, \ldots, a_n$, and it is often denoted by $(a_1, \ldots, a_n)$.

The Division Algorithm has a nice application to the structure of ideals of $\mathbb{Z}$ or of $F[x]$. We prove the result for polynomial rings, leaving the analogous result for $\mathbb{Z}$ to the exercises. An ideal with a single generator is called a principal ideal and integral domains with the property that each of their ideals can be generated by a single element are called *principal ideal domains*.

Theorem 5.11. *Let F be a field. Then $F[x]$ is a principal ideal domain. That is, if I is an ideal of $F[x]$, then there is a polynomial f with $I = (f) = \{fg : g \in F[x]\}$.*

Proof. Let I be an ideal of $F[x]$. If $I = \{0\}$, then $I = (0)$. Suppose that I contains a nonzero element. By the well-ordering property of the integers (applied to the degrees of the nonzero elements of I), there is a nonzero polynomial f in I of least degree. We claim that $I = (f)$. To prove this, let $g \in I$. By the Division Algorithm, there are polynomials q, r with $g = qf + r$ and $\deg(r) < \deg(f)$. Since $f \in I$, the product $qf \in I$, and thus $g - qf \in I$ as $g \in I$. We conclude that $r \in I$. However, the assumption on the degree of f shows that the condition $\deg(r) < \deg(f)$ forces $r = 0$. Thus, $g = qf \in (f)$. This proves $I \subseteq (f)$. Since every multiple of f is in I, the reverse inclusion $(f) \subseteq I$ is also true. Therefore, $I = (f)$. □

The condition that all ideals are principal is very restrictive as the next example shows.

Example 5.12. If F is a field and x, y indeterminates, then $F[x, y]$ is an integral domain but not a principal ideal domain; in particular, the ideal (x, y) cannot be generated by a single element. To see this, suppose to the contrary that $(x, y) = (f)$. Then there are $g, h \in F[x, y]$ with $x = fg$ and $y = fh$. The first equation says that the y degree of f must equal 0 while the second says that its x degree is 0. Thus f is constant and nonzero so that $f \in F$ and $(f) = F[x, y]$. But $(x, y) \neq F[x, y]$; for instance, $1 \notin (x, y)$. Otherwise $1 = xA + yB$, but the right side evaluates to 0 at $x = y = 0$, while the left side doesn't.

We can give an ideal theoretic description of greatest common divisors in $\mathbb{Z}$ and in $F[x]$. Suppose that f and g are polynomials over a field F. If $\gcd(f, g) = d$, then we have proved that $d = fh + gk$ for some polynomials h, k; in particular d is an element of the ideal $I = \{fs + gt : s, t \in F[x]\}$. On the other hand, d divides f and g, so that d divides every element of I and we conclude that $I = (d)$, the principal ideal generated by $\gcd(f, g)$. Therefore, one can identify the greatest common divisor of f and g by identifying a monic polynomial d satisfying $I = (d)$.

We now use ideals to define quotient rings. The construction of a quotient ring is reminiscent of the construction used earlier in the coset decoding algorithm for certain codes. In that context cosets arose from a subspace (namely the code) of a vector space ($\mathbb{Z}_2^n$). The idea here is essentially the same except that the vector space is replaced by a ring and the subspace by an ideal of the ring.

Definition 5.13. Let R be a ring and let I be an ideal of R. If $a \in R$, then the coset $a + I$ is defined as $a + I = \{a + x : x \in I\}$.

Recall the description of equivalence classes for the relation congruence modulo n. For example, if $n = 5$, then we have five equivalence classes, and they are

$$\overline{0} = \{0, 5, 10, \ldots, -5, -10, \ldots\},$$
$$\overline{1} = \{1, 6, 11, \ldots, -4, -9, -14, \ldots\},$$
$$\overline{2} = \{2, 7, 12, \ldots, -3, -8, -13, \ldots\},$$
$$\overline{3} = \{3, 8, 13, \ldots, -2, -7, -12, \ldots\},$$
$$\overline{4} = \{4, 9, 14, \ldots, -1, -6, -11, \ldots\}.$$

By the first example above, the set $5\mathbb{Z}$ of multiples of 5 forms an ideal of $\mathbb{Z}$. These five equivalence classes can be described as cosets, namely,

$$\overline{0} = 0 + 5\mathbb{Z},$$
$$\overline{1} = 1 + 5\mathbb{Z},$$
$$\overline{2} = 2 + 5\mathbb{Z},$$
$$\overline{3} = 3 + 5\mathbb{Z},$$
$$\overline{4} = 4 + 5\mathbb{Z}.$$

In general, for any integer a, we have $a + 5\mathbb{Z} = \overline{a}$. Thus, cosets for the ideal $5\mathbb{Z}$ are the same as equivalence classes modulo 5. In fact, more generally, if n is any positive integer, then the equivalence class $\overline{a}$ of an integer a modulo n is the coset $a + n\mathbb{Z}$ of the ideal $n\mathbb{Z}$.

We have seen that the same equivalence class can have different names. Modulo 5, we have $\overline{1} = \overline{6}$ and $\overline{2} = \overline{-3} = \overline{22}$, for example. Similarly, cosets can be represented in different ways. If $R = F[x]$ and $I = xR$, the ideal of multiples of the polynomial x, then $0 + I = x + I = x^2 + I = 4x^{17} + I$. Also, $1 + I = (x + 1) + I$. For some terminology, we refer to a as a *coset representative* of $a + I$. One important thing to remember is that the coset representative is not unique, as the examples above demonstrate.

When we defined operations on $\mathbb{Z}_n$, we defined them with the formulas $\overline{a} + \overline{b} = \overline{a + b}$ and $\overline{a} \cdot \overline{b} = \overline{ab}$. Since these equivalence classes are the same thing as cosets for $n\mathbb{Z}$, this leads us to consider a generalization. If we replace $\mathbb{Z}$ by any ring and $n\mathbb{Z}$ by any ideal, we can mimic these formulas to define operations on cosets. First, we establish a notation for the set of cosets.

Definition 5.14. If I is an ideal of a ring R, let R/I denote the set of cosets of I. In other words, $R/I = \{a + I : a \in R\}$.

We now define operations on R/I in a manner like the operations on $\mathbb{Z}_n$. We define

$$(a + I) + (b + I) = (a + b) + I,$$
$$(a + I) \cdot (b + I) = (ab) + I.$$

In other words, to add or multiply two cosets, first add or multiply their coset representatives, then take the corresponding coset. As with the operations on $\mathbb{Z}_n$, we have to check that these formulas make sense. In other words, the name we give to a coset should not affect the value we get when adding or multiplying. We first need to know when two elements represent the same coset. To help with the proof, we point out two simple properties. If I is an ideal, then $0 \in I$. Furthermore, if $r \in I$, then $-r \in I$. The proofs of these facts are left as exercises.

Lemma 5.15. *Let I be an ideal of a ring R. If $a, b \in R$, then $a + I = b + I$ if and only if $a - b \in I$.*

Proof. Let $a, b \in R$. First suppose that $a + I = b + I$. From $0 \in I$ we get $a = a + 0 \in a + I$, so $a \in b + I$. Therefore, there is an $x \in I$ with $a = b + x$. Thus, $a - b = x \in I$. Conversely, suppose that $a - b \in I$. If we set $x = a - b$, an element of I, then $a = b + x$. This shows $a \in b + I$. So, for any $y \in I$, we have $a + y = b + (x + y) \in I$, as I is closed under addition. Therefore, $a + I \subseteq b + I$. The reverse inclusion is similar; by using $-x = b - a$, again an element of I, we will get the inclusion $b + I \subseteq a + I$, and so $a + I = b + I$. □

In fact, we can generalize the fact that equivalence classes modulo n are the same thing as cosets for $n\mathbb{Z}$. Given an ideal, we can define an equivalence relation by mimicking congruence modulo n. To phrase this relation in a new way, $a \equiv b \bmod n$ if and only if $a - b$ is a multiple of n, so $a \equiv b \bmod n$ if and only if $a - b \in n\mathbb{Z}$. Thus, given an ideal I of a ring R, we may define a relation by $x \equiv y \bmod I$ if $x - y \in I$. One can prove in the same manner as for congruence modulo n that this is an equivalence relation, and that, for any $a \in R$, the coset $a + I$ is the equivalence class of a.

Lemma 5.16. *Let I be an ideal of a ring R. Let $a, b, c, d \in R$.*

1. *If $a + I = c + I$ and $b + I = d + I$, then $(a + b) + I = (c + d) + I$.*
2. *If $a + I = c + I$ and $b + I = d + I$, then $ab + I = cd + I$.*

Proof. Suppose that $a, b, c, d \in R$ satisfy $a + I = c + I$ and $b + I = d + I$. To prove the first statement, by the lemma we have elements $x, y \in I$ with $a - c = x$ and $b - d = y$. Then

$$\begin{aligned}(a + b) - (c + d) &= a + b - c - d \\ &= (a - c) + (b - d) \\ &= x + y \in I.\end{aligned}$$

Therefore, again by the lemma, $(a+b)+I = (c+d)+I$. For the second statement, we rewrite the equations above as $a = c + x$ and $b = d + y$. Then

$$\begin{aligned}ab = (c + x)(d + y) &= c(d + y) + x(d + y) \\ &= cd + (cy + xd + xy).\end{aligned}$$

Since $x, y \in I$, the three elements cy, xd, xy are all elements of I. Thus, the sum $cy+xd+xy \in I$. This shows us that $ab - cd \in I$, so the lemma yields $ab + I = cd + I$. □

The consequence of the lemma is exactly that our coset operations are well defined. Thus, we can ask whether or not R/I is a ring. The answer is yes, and the proof is easy, and is exactly parallel to the proof for $\mathbb{Z}_n$.

Theorem 5.17. *Let I be an ideal of a ring R. Then R/I, together with the operations of coset addition and multiplication, forms a ring.*

Proof. We have several properties to verify. Most follow immediately from the definition of the operations and from the ring properties of R. For example, to prove that coset addition is commutative, we see that for any $a, b \in R$, we have

$$\begin{aligned}(a + I) + (b + I) &= (a + b) + I \\ &= (b + a) + I \\ &= (b + I) + (a + I).\end{aligned}$$

This used exactly the definition of coset addition and commutativity of addition in R. Most of the other ring properties hold for similar reasons, so we only verify those that are a little different.

For the existence of an additive identity, it is natural to guess that $0 + I$ is the identity for R/I where 0 is the additive identity of R. To see that this guess is correct, let $a + I \in R/I$. Then

$$(a + I) + (0 + I) = (a + 0) + I = a + I.$$

Thus, $0 + I$ is the additive identity for R/I. Similarly $1 + I$ is the multiplicative identity, since

$$(a + I) \cdot (1 + I) = (a \cdot 1) + I = a + I$$

and

$$(1 + I) \cdot (a + I) = (1 \cdot a) + I = a + I$$

for all $a + I \in R/I$. Finally, the additive inverse of $a + I$ is $-a + I$ since

$$(a + I) + (-a + I) = (a + (-a)) + I = 0 + I.$$

Therefore, R/I is a ring. □

The ring R/I is called a *quotient ring* of R. This idea allows us to construct new rings from old rings. For example, the ring $\mathbb{Z}_n$ is really the same thing as the quotient ring $\mathbb{Z}/n\mathbb{Z}$, since we have identified the equivalence classes modulo n; that is, the elements of $\mathbb{Z}_n$, with the cosets of $n\mathbb{Z}$; i.e., the elements of $\mathbb{Z}/n\mathbb{Z}$. It is this construction applied to polynomial rings that we will use to build extension fields. We recall Proposition 3.29 above that says $\mathbb{Z}_n$ is a field if and only if n is a prime. To generalize this result to polynomials, we first need to define the polynomial analogue of a prime number.

Definition 5.18. Let R be a commutative ring. An element $r \in R$ is said to be irreducible if $r = ab$ implies that either a is a unit or b is a unit.

The case of irreducible elements in polynomial ring over a field is of such importance that we isolate that case in the following definition.

Definition 5.19. Let F be a field. A nonconstant polynomial $f \in F[x]$ is said to be irreducible over F if whenever f can be factored as $f = gh$, then either g or h is a constant polynomial.

Note that a polynomial $f \in F[x]$ that is irreducible in this context is irreducible as an element of the ring $F[x]$ since a constant polynomial is simply a polynomial of degree 0; that is, it is a polynomial of the form $f(x) = a$ for some $a \in F$. Any such polynomial has degree 0 and is a unit if it is not the zero polynomial.

Example 5.20. The terminology irreducible over F in the definition above is used because irreducibility is a relative term. The polynomial x^2+1 factors over $\mathbb{C}$ as $x^2+1 = (x-i)(x+i)$.

In contrast, x^2+1 is irreducible over $\mathbb{R}$. To see this, one could write $x^2+1=(ax+b)(cx+d)$ and collect coefficients of $1, x$, and x^2 to obtain a system of nonlinear equations in a, b, c, d, for which there is no solution in real numbers. An easier method is as follows: Since $\deg(x^2+1)=2$, if it factors over $\mathbb{R}$, then it must have a root in $\mathbb{R}$ (Exercise 8 of Sect. 5.1). However, squares of real numbers are positive, so x^2+1 has no roots in $\mathbb{R}$ and is therefore irreducible over $\mathbb{R}$.

Example 5.21. The polynomial x is irreducible: Given a factorization $x = gh$, taking degrees of both sides gives $1 = \deg(g) + \deg(h)$. Thus, one of the degrees of g and h is 1 and the other is 0. The one with degree 0 is a constant polynomial. Since we cannot factor x with both factors nonconstant, x is irreducible. This argument shows that every polynomial of degree 1 is irreducible.

Example 5.22. Consider x^2+1 as a polynomial in $\mathbb{Z}_5[x]$. Unlike the case of $\mathbb{Q}[x]$, this polynomial does factor over $\mathbb{Z}_5$, since $x^2+1=(x-2)(x-3)$ in $\mathbb{Z}_5[x]$. In particular, x^2+1 has two roots in $\mathbb{Z}_5$. However, for $F=\mathbb{Z}_3$, the polynomial x^2+1 is irreducible since x^2+1 has no roots in $\mathbb{Z}_3$; it is easy to see that none of the three elements 0, 1, and 2 are roots of x^2+1.

To help work with quotient rings $F[x]/I$, we can use the Division Algorithm to write elements in a normalized form. Set $I=(f)$. Given $g \in F[x]$, by the Division Algorithm we may write $g = qf + r$ for some $g, r \in F[x]$ and with $\deg(r) < \deg(f)$. Then $g - r = qf \in I$, so $g + I = r + I$. This argument shows that any coset $g + I$ is equal to a coset $r + I$ for some polynomial r with $\deg(r) < \deg(f)$. Thus, $F[x]/(f) = \{r + I : r \in F[x], \deg(r) < \deg(f)\}$. This result is the analogue of the description $\mathbb{Z}_n = \{\overline{a} : 0 \le a < n\} = \{a + (n) : 0 \le a < n\}$.

Let $F=\mathbb{R}$, and consider the irreducible polynomial $f = x^2+1$. In this example we will use the normalized form of elements to relate the field $\mathbb{R}[x]/(x^2+1)$ to the field of complex numbers $\mathbb{C}$. The Division Algorithm implies that every element of this quotient ring can be written in the form $a + bx + I$, where $I = (x^2+1)$. Addition in this ring is given by

$$(a + bx + I) + (c + dx + I) = (a + c) + (b + d)x + I.$$

For multiplication, we have

$$\begin{aligned}(a + bx + I)(c + dx + I) &= (a + bx)(c + dx) + I \\ &= ac + bdx^2 + (ad + bc)x + I \\ &= (ac - bd) + (ad + bc)x + I;\end{aligned}$$

the simplification in the last equation comes from the equation $bdx^2+I = -bd+I$. Since $x^2+1 \in I$, we have $x^2 + I = -1 + I$, so multiplying both sides by $bd + I$ yields this equation. If you look at these formulas for the operations in $\mathbb{R}[x]/(x^2+1)$, you may see a similarity between the operations on $\mathbb{C}$:

$$\begin{aligned}(a + bi) + (c + di) &= (a + c) + (b + d)i \\ (a + bi)(c + di) &= (ac - bd) + (ad + bc)i.\end{aligned}$$

In fact, one can view this construction as a way of building the complex numbers from the real numbers and the polynomial x^2+1.

Exercises

1. Prove that every ideal in the ring $\mathbb{Z}$ of integers can be generated by a single element. (Hint: Follow the proof of Theorem 5.11 and replace "nonzero polynomial f in I of least degree" by "nonzero element of least absolute value.")
2. Prove that the ideal $(2, x)$ in the ring $\mathbb{Z}[x]$ cannot be generated by a single element.
3. Prove that if R is an integral domain then so is $R[x]$ for an indeterminate x. Conclude for every finite collection $\{x_1, x_2, \dots x_n\}$ of indeterminates that $R[x_1, x_2, \dots x_n]$ is an integral domain.
4. Let F be a field, let $f \in F[x]$, and let $I = (f)$ be the ideal generated by f.

   (a) Show that, for any $g \in F[x]$, if $g = qf + r$ with $\deg(r) < \deg(f)$, then $g + I = r + I$. Conclude that any coset $g + I$ can be represented by a polynomial of degree less than $\deg(f)$.
   (b) If $\deg(g) < \deg(f)$ and $\deg(h) < \deg(f)$, show that $g + I = h + I$ if and only if $g = h$.

5. Determine all of the cosets of the ideal $(x^2 + x + 1)$ in the ring $\mathbb{Z}_2[x]$. Instead of writing out all the elements in a coset (for which there are infinitely many), just find a coset representative for each coset and say why you have produced all cosets.
   (Hint: Use the previous problem.)
6. Determine all of the cosets of the ideal $(x^3 + x + 1)$ in the ring $\mathbb{Z}_2[x]$.
7. Write out the elements of F as cosets of polynomials of degree at most 2, and then write out the addition and multiplication tables for F. Feel free to take advantage of the commands plus and mult defined in the worksheet.
8. Write out the powers $(\overline{x})^i$ for $1 \leq i \leq 7$ as cosets of polynomials of degree at most 2. Feel free to use the powers command defined in the worksheet.
9. Express the following cosets as cosets of polynomials of degree at most 2.

   (a) $\overline{x^8}$
   (b) $\overline{x^5 + x + 1}$
   (c) $\overline{x^4 + x^3 + x^2}$
   (d) $\overline{x^6 + x}$.

5.3 Field Extensions

We now see how the last example of the previous section generalizes to enable the construction of field extensions from irreducible polynomials. Recall an earlier definition.

Definition 5.23. If F is a field, then a *field extension of* F is a field K that contains F, and for which the field operations restrict to the field operation of F. Under these circumstances we also say that F is a subfield of K.

Proposition 5.24. *Let F be a field, and let $f \in F[x]$ be a polynomial. If $I = (f)$ is the ideal generated by an irreducible polynomial f, then $F[x]/I$ is a field extension of F.*

Proof. Let F be a field, and let $f \in F[x]$ be irreducible. Set $I = (f)$. We wish to prove that $F[x]/I$ is a field. We know it is a commutative ring, so we only need to prove that every nonzero element has a multiplicative inverse. Let $g + I \in F[x]/I$ be nonzero. Then $g + I \neq 0 + I$, so $g \notin I$. This means f does not divide g. Since f is irreducible, we can conclude that $\gcd(f, g) = 1$. Thus, there are $h, k \in F[x]$ with $1 = hf + kg$. Because $hf \in I$, $kg - 1 \in I$, so $kg + I = 1 + I$. By the definition of coset multiplication, this yields $(k + I)(g + I) = 1 + I$. Therefore, $k + I$ is the multiplicative inverse of $g + I$. Because we have proved that an arbitrary nonzero element of $F[x]/I$ has a multiplicative inverse, this commutative ring is a field.

The elements $a + I$ with $a \in F$ clearly form a subfield of $F[x]/I$, which is identified with the field F. In this way $F[x]/I$ is realized as a field extension of F. □

The converse of this result is also true; if $F[x]/(f)$ is a field, then f is an irreducible polynomial. We leave the verification of a slightly more general statement to Exercise 11 below.

Example 5.25. Let $F = \mathbb{Q}$ and $f = x^3 - 2$. The Rational Roots Test shows that f has no roots in $\mathbb{Q}$ and is therefore irreducible over that field. Thus $\mathbb{Q}[x]/(x^3 - 2)$ is a field. Moreover, the normalization representation of elements of $\mathbb{Q}[x]/(x^3 - 2)$ shows that each element can be written as $a \cdot 1 + b \cdot x + c \cdot x^2 + (x^3 - 2)$ for certain $a, b, c \in \mathbb{Q}$. This representation is unique since $a \cdot 1 + b \cdot x + c \cdot x^2 + (x^3 - 2) = a' \cdot 1 + b' \cdot x + c' \cdot x^2 + (x^3 - 2)$ implies that

$$(a - a') \cdot 1 + (b - b') \cdot x + (c - c' \cdot)x^2 + (x^3 - 2) = 0 + (x^3 - 2),$$

i.e., that $x^3 - 2$ divides the polynomial $(a - a') \cdot 1 + (b - b') \cdot x + (c - c' \cdot)x^2$. But by considering degrees, this is impossible unless the latter polynomial is 0.

Example 5.26. Let $F = \mathbb{Z}_2$, and consider $f = x^2 + x + 1$. This is irreducible over $\mathbb{Z}_2$ since it is quadratic and has no roots in $\mathbb{Z}_2$; the only elements of $\mathbb{Z}_2$ are 0 and 1, and neither is a root. Consider $K = \mathbb{Z}_2[x]/(x^2 + x + 1)$. This is a field by the previous proposition. We write out an addition and multiplication table for K once we write down all elements of K. First, by the comment above, any coset in K can be represented by a polynomial of the form $ax + b$ with $a, b \in \mathbb{Z}_2$; this is because any remainder after division by f must have degree less than $\deg(f) = 2$. So,

$$K = \{0 + I, 1 + I, x + I, 1 + x + I\}.$$

Thus, K is a field with 4 elements. The following tables then represent addition and multiplication in K.

$+$	$0 + I$	$1 + I$	$x + I$	$1 + x + I$
$0 + I$	$0 + I$	$1 + I$	$x + I$	$1 + x + I$
$1 + I$	$1 + I$	$0 + I$	$1 + x + I$	$x + I$
$x + I$	$x + I$	$1 + x + I$	$0 + I$	$1 + I$
$1 + x + I$	$1 + x + I$	$x + I$	$1 + I$	$0 + I$

$\cdot$	$0 + I$	$1 + I$	$x + I$	$1 + x + I$
$0 + I$	$0 + I$	$0 + I$	$0 + I$	$0 + I$
$1 + I$	$0 + I$	$1 + I$	$x + I$	$1 + x + I$
$x + I$	$0 + I$	$x + I$	$1 + x + I$	$1 + I$
$1 + x + I$	$0 + I$	$1 + x + I$	$1 + I$	$1 + x + I$

If you look closely at these tables, you may see a resemblance between them and the tables of Example 3.30 above. In fact, if you label $x + I$ as a and $1 + x + I$ as b, along with $0 = 0 + I$ and $1 = 1 + I$, the tables in both cases are identical. In fact, the tables of Example 3.30 were found by building K, and then labeling the elements as 0, 1, a, and b in place of $0 + I$, $1 + I$, $x + I$, and $1 + x + I$.

Example 5.27. Let $f(x) = x^3 + x + 1$. Then this polynomial is irreducible over $\mathbb{Z}_2$. To see this, we first note that it has no roots in $\mathbb{Z}_2$ as $f(0) = f(1) = 1$. Since $\deg(f) = 3$, if it factored, then it would have a linear factor, and so a root in $\mathbb{Z}_2$. Since this does not happen, it is irreducible. We consider the field $K = \mathbb{Z}_2[x]/(x^3 + x + 1)$. We write $I = (x^3 + x + 1)$ and set $\alpha = x + I$. We first note an interesting fact about this field; every nonzero element of K is a power of α. First of all, we have

$$K = \{0 + I, 1 + I, x + I, (x + 1) + I, x^2 + I, (x^2 + 1) + I, (x^2 + x) + I, (x^2 + x + 1) + I\}.$$

We then see that

$$\begin{aligned}
\alpha &= x + I, \\
\alpha^2 &= x^2 + I, \\
\alpha^3 &= x^3 + I = (x + 1) + I = \alpha + 1 \\
\alpha^4 &= x^4 + I = x(x + 1) + I = (x^2 + x) + I = \alpha^2 + \alpha \\
\alpha^5 &= x^5 + I = x^3 + x^2 + I = (x^2 + x + 1) + I = \alpha^2 + \alpha + 1 \\
\alpha^6 &= x^6 + I = (\alpha^3)^2 = (x^2 + 1) + I = \alpha^2 + 1 \\
\alpha^7 &= x^7 + I = x(x^2 + 1) + I = x^3 + x + I = 1 + I.
\end{aligned}$$

To obtain these equations we took several steps. For example, we used the definition of coset multiplication. For instance, $\alpha^2 = (x + I)^2 = x^2 + I$ from this definition. Next, for $\alpha^3 = x^3 + I$, since $x^3 + x + 1 \in I$, we have $x^3 + I = (x + 1) + I$. For other equations, we used combinations of these ideas. For example, to simplify $\alpha^5 = x^5 + I$, first note that

$$\begin{aligned}
\alpha^5 = \alpha^3 \cdot \alpha^2 &= ((x + 1) + I)\left(x^2 + I\right) \\
&= x^3 + x^2 + I \\
&= (x^2 + x + 1) + I
\end{aligned}$$

since $x^3 + x + 1 \in I$.

Familiar concepts from linear algebra (i.e., linear independence, spanning, basis, and dimension) provide essential tools for the study of field extensions. For instance, if K is an extension field of F, then K has the structure of an F-vector space with the field addition in K as its addition and scalar multiplication $\alpha\beta$ for $\alpha \in F, \beta \in K$ given by the multiplication in the field K (since F is a subfield of K this makes sense). A proof of this basic fact is Exercise 1 of this section.

Example 5.28. Let $F = \mathbb{Q}$ and $K = \{\alpha + \beta\sqrt{2} : \alpha, \beta \in \mathbb{Q}\}$. Since

$$(\alpha_1 + \beta_1\sqrt{2})(\alpha_2 + \beta_2\sqrt{2}) = (\alpha_1\alpha_2 + 2\beta_1\beta_2) + (\alpha_1\beta_2 + \beta_1\alpha_2)\sqrt{2},$$

we see that K is closed under multiplication. It is clearly closed under addition and contains $\mathbb{Q}$ as the subset $\{\alpha + \beta\sqrt{2} : \alpha \in \mathbb{Q}, \beta = 0\}$. Note that if either $\alpha \neq 0$ or $\beta \neq 0$ then $\alpha^2 + 2\beta^2 \neq 0$. A multiplication reveals that

$$\left(\frac{\alpha}{\alpha^2 + 2\beta^2} - \frac{\beta}{\alpha^2 + 2\beta^2}\sqrt{2}\right)(\alpha + \beta\sqrt{2}) = 1$$

so that $\frac{\alpha}{\alpha^2+2\beta^2} - \frac{\beta}{\alpha^2+2\beta^2}\sqrt{2}$ is the multiplicative inverse of $\alpha + \beta\sqrt{2}$ and is an element of K. Thus nonzero elements of K all have multiplicative inverses.

Definition 5.29. Let K be a field extension of F. We denote the dimension of K as an F-vector space by $[K : F]$.

Example 5.30. The dimension of $\mathbb{Q}(\sqrt{2})$ as a $\mathbb{Q}$-vector space is 2 since 1 and $\sqrt{2}$ form a basis. Thus, $\left[\mathbb{Q}\left(\sqrt{2}\right) : \mathbb{Q}\right] = 2$.

Given a field extension K of F and an element $\alpha \in K$ we denote by $F(\alpha)$ the smallest field extension of F which contains α. By smallest we mean that for any other field extension L of F with $\alpha \in L$, we have $F(\alpha) \subseteq L$. For $\alpha, \beta \in K$ we write $F(\alpha, \beta)$ instead of $F(\alpha)(\beta)$. Similarly for any finite collection $\{\alpha_1, \alpha_2, \dots, \alpha_n\}$ of elements of K we write $F(\alpha_1, \alpha_2, \dots, \alpha_n)$ for the smallest field extension of F containing these elements.

Example 5.31. Let $K = \{\alpha + \beta\sqrt{2} : \alpha, \beta \in \mathbb{Q}\}$ as above. Since every field extension of $\mathbb{Q}$ containing $\sqrt{2}$ must contain every real number of the form $\alpha + \beta\sqrt{2}$ with $\alpha, \beta \in \mathbb{Q}$, and K is a field, $K = \mathbb{Q}(\sqrt{2})$.

Example 5.32. With i denoting a complex square root of -1, $\mathbb{R}(i) = \mathbb{C}$. Because every complex number is represented uniquely as $a + bi$ with $a, b \in \mathbb{R}$, the complex numbers $\{1, i\}$ form a basis for $\mathbb{C}$ as an $\mathbb{R}$-vector space and $[\mathbb{C} : \mathbb{R}] = 2$

Example 5.33. The cube roots of 1 solve $x^3 - 1 = 0$. Since $x^3 - 1 = (x-1)(x^2 + x + 1)$, the three cube roots of 1 are

$$1,\ \omega_1 = \frac{-1 + i\sqrt{3}}{2},\ \text{and}\ \omega_2 = \frac{-1 - i\sqrt{3}}{2},$$

with ω_1 and ω_2 complex conjugates, in particular they do not lie in $\mathbb{R}$. In Exercise 6 you will show that $\mathbb{Q}(\omega_1)$ contains all three cube roots of 1 and Example 5.48 shows that $[\mathbb{Q}(\omega_1) : \mathbb{Q}] = 2$.

Proposition 5.34. *Let F be a field and $f \in F[x]$ an irreducible polynomial. Then $[F[x]/(f) : F] = \deg(f)$.*

Proof. Let $n = \deg(f)$ and to simplify notations, write elements $a + (f)$ with $a \in F$ simply as a and $x^i + (f)$ as $\overline{x}^i$. Then the normalized form of any element of $F[x]/(f)$ is $\sum_{i=0}^{n-1} a_i\overline{x}^i$ for suitable $a_i \in F$, so we see that $\{1, \overline{x}, \overline{x}^2, \dots, \overline{x}^{n-1}\}$ spans $F[x]/(f)$ as an F vector space. This set is also linearly independent. Indeed, if $\sum_{i=0}^{n-1} a_i\overline{x}^i = 0$, with $a_i \in F$, then $\sum_{i=0}^{n-1} a_i x^i \in (f)$, i.e., the polynomial f, which has degree n, divides the lower degree polynomial $\sum_{i=0}^{n-1} a_i x^i$. This implies that the latter polynomial is the 0 polynomial, i.e., that $a_i = 0$ for each i. □

We often encounter a tower of fields $F \subseteq K \subseteq L$. For example, with ω_1 ω_2 as in Example 5.33, the three roots of $x^3 - 2 \in \mathbb{Q}[x]$ are $\sqrt[3]{2}$, $\sqrt[3]{2}\omega_1$, and $\sqrt[3]{2}\omega_2$ so that all three roots lie in $\mathbb{Q}(\sqrt[3]{2}, \omega_1)$ (refer again to Exercise 6 below). Since $\sqrt[3]{2} \in \mathbb{R}$ but $\omega_1, \omega_2 \notin \mathbb{R}$, we have the tower $\mathbb{Q} \subset \mathbb{Q}(\sqrt[3]{2}) \subset \mathbb{Q}(\sqrt[3]{2}, \omega_1)$.

Theorem 5.35 (Dimension Formula). *Let K be a field extension of F and L a field extension of K. Then*

$$[L : F] = [L : K][K : F].$$

Proof. If either $[L : K]$ or $[K : F]$ is infinite, then so is $[L : F]$ and there is nothing to show. So suppose that $[K : F] = m$ and $[L : K] = n$. Let $\{a_1, \ldots, a_m\}$ and $\{b_1, \ldots, b_n\}$ be bases for K over F and L over K, respectively. We show that the elements $a_i b_j \in L$, for $1 \leq i \leq m$, $1 \leq j \leq n$, constitute a basis for L as an F vector space.

For linear independence, suppose that $\sum_{1 \leq i \leq m, 1 \leq j \leq n} \gamma_{ij} a_i b_j = 0$ for some $\gamma_{ij} \in F$. Then

$$\sum_{1 \leq j \leq n} \left(\sum_{1 \leq i \leq m} \gamma_{ij} a_i \right) b_j = 0$$

gives a dependence relation among $b_1, \ldots, b_n$ over K. Since they are linearly independent, we must have the coefficient of each b_j equal to 0, i.e.,

$$\sum_{1 \leq i \leq m} \gamma_{ij} a_i = 0 \text{ for each } j.$$

But the linear independence of the a_i then forces each $\gamma_{ij} = 0$.

To see that $\{a_i b_j : 1 \leq i \leq m, 1 \leq j \leq n\}$ spans L over F, let $c \in L$. Since $\{b_1, \ldots, b_n\}$ spans L over K we can write $c = \sum_{1 \leq j \leq n} \alpha_j b_j$ for some $\alpha_j \in K$. Each α_j can be expanded as an F linear combination of the a_i as

$$\alpha_j = \sum_{1 \leq i \leq m} \beta_{ij} a_i.$$

Combining these expressions we obtain

$$c = \sum_{1 \leq j \leq n} \alpha_j b_j = \sum_{1 \leq j \leq n} \sum_{1 \leq i \leq m} \beta_{ij} a_i b_j.$$

□

Example 5.36. The polynomial $x^3 - 2 \in \mathbb{Q}[x]$ has the three distinct roots $\sqrt[3]{2}, \sqrt[3]{2}\omega, \sqrt[3]{2}\omega^2$ where $\omega = \omega_1$ of Example 5.33 (see also Exercise 6). Clearly all three roots of $x^3 - 2$ lie in $\mathbb{Q}(\sqrt[3]{2}, \omega)$ and conversely every field which contains these roots must also contain $\mathbb{Q}(\sqrt[3]{2}, \omega)$. Therefore $\mathbb{Q}(\sqrt[3]{2}, \omega)$ is the smallest field extension of $\mathbb{Q}$ containing all roots of $x^3 - 2$. Since $\mathbb{Q}(\sqrt[3]{2}) \subset \mathbb{R}$ and $\omega \notin \mathbb{R}$, we have the tower of fields $\mathbb{Q} \subset \mathbb{Q}(\sqrt[3]{2}) \subset \mathbb{Q}(\sqrt[3]{2}, \omega)$ and both inclusions are proper. In Exercise 6 you will show that $[\mathbb{Q}(\omega) : \mathbb{Q}] = 2$ so that $[\mathbb{Q}(\sqrt[3]{2}, \omega) : \mathbb{Q}] = 2[\mathbb{Q}(\sqrt[3]{2} : \mathbb{Q}]$. In the next section you will see that $[\mathbb{Q}(\sqrt[3]{2} : \mathbb{Q}] = 3$ so that $[\mathbb{Q}(\sqrt[3]{2}, \omega) : \mathbb{Q}] = 6$. In other words it is necessary to extend to an extension of degree 6 to obtain the smallest field extension of $\mathbb{Q}$ containing all three roots of this cubic polynomial.

5.4 Algebraic Elements and Minimal Polynomials

We conclude this chapter with an important concept that connects the field extensions $F[x]/(f)$ and $F(\alpha)$ of F where $f \in F[x]$ is irreducible and α is a root of f. The assertion that α is a root of f means that there is some field extension K of F containing α and that as an element of K, the evaluation of $f(\alpha)$ yields 0. In other words, although the coefficients of f lie in F, it can be viewed as having its coefficients in K so that $f(\alpha) = 0$ is meaningful computed in the field K.

Lemma 5.37. *Let K be an extension field of F and $\alpha \in K$. Then*

$$I = \{g \in F[x] : g(\alpha) = 0\}$$

is an ideal of $F[x]$. If I is not the 0 ideal, then $I = (f)$ for some irreducible $f \in F[x]$.

Proof. That I is an ideal is an easy exercise. Suppose that $I \neq (0)$ so that $I = (f)$ for some $f \in I$ of positive degree, indeed we can assume that the degree of f is minimal among all nonzero elements of I. If $f = gh$ with $g, h \in F[x]$, then $0 = f(\alpha) = (gh)(\alpha) = g(\alpha)h(\alpha)$. Since this computation takes place in the field K, one of $g(\alpha) = 0$ or $h(\alpha) = 0$ must hold, i.e., $g \in I$ or $h \in I$. Since the degree of f is minimal among all nonzero elements of I, it must be that g or h has degree 0, i.e., that g or h is a unit. □

Definition 5.38. Let K be an extension field of F, $\alpha \in K$, and $I = \{g \in F[x] : g(\alpha) = 0\}$. If $I \neq (0)$, then α is said to be algebraic over F. If $I = (0)$ then α is said to be transcendental over F.

In simpler terms α is algebraic over F if α is a root of some polynomial of positive degree with coefficients in F.

Example 5.39. Taking $F = \mathbb{R}$, $K = \mathbb{C}$, and $\alpha = i = \sqrt{-1}$, we see that $I = (x^2 - 1)$ so that i is algebraic over $\mathbb{R}$.

Example 5.40. Taking $F = \mathbb{Q}$, $K = \mathbb{R}$, and $\alpha = \pi$, an important and difficult theorem asserts that π is transcendental over $\mathbb{Q}$.

Given a field extension K of the field F, suppose that $\alpha \in K$ is algebraic over F. Let f be a generator of $I = \{g \in F[x] : g(\alpha) = 0\}$, written as $f = \sum_{i=0}^{n} a_i x^i$ with $a_i \in F$ and $a_n \neq 0$. Then $a_n^{-1} f$ is monic of the same degree as f. Since f has least positive degree among the nonzero elements of I, the polynomial $a_n^{-1} f$ is the unique monic polynomial in I.

Definition 5.41. With F, K, α, f as above, the polynomial $a_n^{-1} f$ is called the minimal polynomial of α over F.

Notation 5.42. To emphasize the roles of F and α in the definition of minimal polynomial, the symbol $\min_F(\alpha)$ is used to denote it.

Example 5.43. The real numbers $\sqrt{2}$ and $\sqrt[4]{2}$ are clearly algebraic over $\mathbb{Q}$. It is straightforward to show that $\min_{\mathbb{Q}}(\sqrt{2}) = x^2 - 2$ and not very hard to show that $\min_{\mathbb{Q}}(\sqrt[4]{2}) = x^4 - 2$ (Exercise 9). Setting $f = x^2 - \sqrt{2} \in \mathbb{Q}(\sqrt{2})[x]$, and noting that $f(\sqrt[4]{2}) = 0$, we see that $f = \min_{\mathbb{Q}(\sqrt{2})}(\sqrt[4]{2})$ provided this polynomial is irreducible over $\mathbb{Q}(\sqrt{2})$.

Definition 5.44. Let F and K be fields. A homomorphism $\varphi : F \to K$ is a function satisfying $f(1_F) = 1_K$ and, for every $a, b \in F$, both of the following hold:

$$f(a + b) = f(a) + f(b)$$
$$f(ab) = f(a)f(b).$$

A field homomorphism is always one-to-one (Exercise 18). An onto field homomorphism $\varphi : F \to K$ is called an isomorphism and F and K are said to be isomorphic.

Theorem 5.45. *Let F be a field, $f \in F[x]$ an irreducible polynomial, and α a root of f. Then there is an isomorphism of fields $F[x]/(f) \to F(\alpha)$. This isomorphism is also an isomorphism of F vector spaces.*

Proof. For $g = \sum_{i=0}^{m} a_i x^i \in F[x]$ define $\varphi(g + (f)) = \sum_{i=0}^{m} a_i \alpha^i = g(\alpha) \in F(\alpha)$. That φ satisfies the three conditions above is a straightforward computation. It is necessary however to prove that φ is well defined and that its image is a field. If the latter is shown, then the image is certainly the smallest field extension of F containing α, namely $F(\alpha)$.

To see that φ is well defined, suppose that $g + (f) = h + (f)$. Then $h = g + fk$ for some $k \in F[x]$ and we have

$$\begin{aligned}\varphi(h + (f)) &= h(\alpha) = g(\alpha) + (fk)(\alpha)\\ &= g(\alpha) + f(\alpha)k(\alpha)\\ &= g(\alpha) = \varphi(g + (f))\end{aligned}$$

because α is a root of f.

That the image of φ is a field (isomorphic to $F[x]/(f)$) follows if it shown that φ is one-to-one. To that end, let $g + (f), h + (f) \in F[x]/(f)$ be written in normalized form, so that $\deg(g), \deg(h) < \deg(f)$. If $\varphi(h + (f)) = \varphi(g + (f))$, then $h(\alpha) = g(\alpha)$ and $(h - g)(\alpha) = 0$. By Lemma 5.37, $h - g \in (f)$ which forces $h - g = 0$, again by consideration of degrees. Thus φ is a field isomorphism. Noting that φ restricts to the identity on the field F we have, for $a \in F$ and $g \in F[x]$, that

$$\begin{aligned}\varphi(ag + (f)) &= \varphi(a)\varphi(g + (f))\\ &= a\varphi(g + (f)),\end{aligned}$$

showing that φ is a vector space isomorphism. □

Corollary 5.46. *Let K be an extension field of F and $\alpha \in K$ algebraic over F. Then $[F(\alpha) : F] = \deg(\min_F(\alpha))$.*

Proof. Since $F(\alpha)$ and $F[x]/(f)$ are isomorphic as F vector spaces, their dimensions over F are equal. □

Example 5.47. The real number $\sqrt[3]{2}$ is certainly a root of the polynomial $x^3 - 2 \in \mathbb{Q}[x]$, and the Rational Roots Test tells us that this polynomial has none. Since its degree is 3 we conclude that $x^3 - 2 = \min_{\mathbb{Q}}(\sqrt[3]{2})$, and that $[\mathbb{Q}(\sqrt[3]{2}) : \mathbb{Q}] = \deg(x^3 - 2) = 3$.

Example 5.48. Let $\omega = \omega_1$ of Example 5.33 and Exercise 6. Since $\mathbb{Q}(\sqrt[3]{2}) \subset \mathbb{R}$ and $\omega_1 \notin \mathbb{R}$, it must be the case that $x^2 + x + 1$ remains irreducible over $\mathbb{Q}(\sqrt[3]{2})$ and therefore $x^2 + x + 1 = \min_{\mathbb{Q}(\sqrt[3]{2})}(\omega)$. In particular $[\mathbb{Q}(\sqrt[3]{2}, \omega) : \mathbb{Q}(\sqrt[3]{2})] = 2$, and $[\mathbb{Q}(\sqrt[3]{2}, \omega_1) : \mathbb{Q}] = [\mathbb{Q}(\sqrt[3]{2}, \omega_1) : \mathbb{Q}(\sqrt[3]{2})][\mathbb{Q}(\sqrt[3]{2}) : \mathbb{Q}] = 6$.

Example 5.49. Since $4 = [\mathbb{Q}(\sqrt[4]{2}) : \mathbb{Q}] = [\mathbb{Q}(\sqrt[4]{2}) : \mathbb{Q}(\sqrt{2})][\mathbb{Q}(\sqrt{2}) : \mathbb{Q}] = 2[\mathbb{Q}(\sqrt[4]{2}) : \mathbb{Q}(\sqrt{2})]$ we can conclude that $[\mathbb{Q}(\sqrt[4]{2}) : \mathbb{Q}(\sqrt{2})]=2$. Therefore $\min_{\mathbb{Q}(\sqrt{2})}(\sqrt[4]{2}) = x^2 - \sqrt{2}$.

Remark 5.50. The image of the isomorphism φ in the Theorem 5.45 is the ring $F[\alpha]$ generated over F by α. But since this ring is a field it must coincide with $F(\alpha)$, which is the smallest field extension of F containing α. The following conditions characterize the algebraic elements α of an extension field K of F. The proof is left to the exercises.

Theorem 5.51. *Let K be an extension field of F and $\alpha \in K$. Then the following conditions are equivalent:*

1) α is algebraic over F
2) the smallest ring extension of F containing α coincides with smallest field extension of F containing α.
3) α is a root of some irreducible polynomial in $F[x]$, and
4) the elements $1, \alpha, \alpha^2, \ldots$ of K are linearly dependent over F.

As a consequence of the Remainder Theorem (Exercise 5 of Sect. 5.1), given $f \in F[x]$ an element α of an extension field K of F is a root of f if and only if $x - \alpha$ divides f in.$K[x]$ (i.e., if and only if $f = (x - \alpha)g$ for some $g \in K[x]$). It follows that all the roots of f lie in F if and only if f factors into linear factors in $F[x]$. More generally, suppose that $n = \deg(f)$, K is an extension field of F, and $f = (x - \alpha)g$ for some $\alpha \in K$ and $g \in K[x]$. Clearly, $\deg(g) = n - 1$, and an induction on n shows that f has at most n roots in K. Viewing $f \in K[x]$ then, f factors into linear factors in $K[x]$ if and only if K contains n roots of f counting multiplicity. For instance, the polynomial $x^4 + 2x^2 + 1 \in \mathbb{Q}[x]$ factors as $(x + i)^2(x - i)^2$ in $\mathbb{Q}(i)[x]$, reflecting the fact that $x^4 + 2x^2 + 1$ has 4 roots in $\mathbb{Q}(i)$, counting each of the roots i and $-i$ with multiplicity 2.

Exercises

1. Given a field extension K of F, prove that the addition in K and product $\alpha\beta$ for $\alpha \in F, \beta \in K$, taken as multiplication in K, endow K with a natural structure of an F vector space.
2. Prove that every field containing the ring of integers as a subring is a field extension of $\mathbb{Q}$.
3. For an integer a set $\mathbb{Q}(\sqrt{a}) = \{\alpha + \beta\sqrt{a} : \alpha, \beta \in \mathbb{Q}\}$. Prove that $\mathbb{Q}(\sqrt{a})$ is a field extension of $\mathbb{Q}$ and $\left[\mathbb{Q}\left(\sqrt{a}\right) : \mathbb{Q}\right] \leq 2$.
4. Let F be a field and $f = ax^2 + bx + c$ a quadratic polynomial in $F[x]$. If α is a root of f, then $[F(\alpha) : F] = 1$ if and only if f is reducible. (Hint: Show that $f = a(x - \alpha)(x - \beta)$ where $\alpha, \beta \in F$).
5. Given a field extension K of F and an element $\alpha \in K$ prove that $F(\alpha)$ is the intersection of all fields L with $F \subseteq L \subseteq K$ and $\alpha \in L$.
6. With $\omega_1 = \frac{-1+i\sqrt{3}}{2}$ and $\omega_2 = \frac{-1-i\sqrt{3}}{2}$ (the complex roots of $x^3 - 1$),verify that $\omega_1 = e^{\frac{2\pi i}{3}}$ and $\omega_2 = \omega_1^2 = e^{\frac{4\pi i}{3}}$, so that $\mathbb{Q}(\omega_1)$ contains all three cube roots of 1.
7. Show for any positive integer n that the set of all n^{th} roots of 1 lies in the field $\mathbb{Q}(e^{\frac{2\pi i}{n}})$.
8. Determine $[\mathbb{Q}(\sqrt{2}, e^{\frac{2\pi i}{3}}) : \mathbb{Q}]$ (Hint: $\mathbb{Q}(\sqrt{2}) \subset \mathbb{R}$.)
9. Prove that $x^4 - 2$ is irreducible over $\mathbb{Q}$ so that $x^4 - 2 = \min_{\mathbb{Q}}(\sqrt[4]{2})$.
10. Let R be an integral domain and x an indeterminate. Prove that $R[x]$ is an integral domain.
11. If $f \in R[x]$ and $R[x]/(f)$ is an integral domain, prove that f is irreducible.
12. For a field extension K of F, prove $K = F$ if and only if $[K : F] = 1$.
13. If K and L are field extensions of F, prove that $K \cap L$ is a field extension of F.
14. If K and L are field extensions of F with $[L : F] = p$ and $[K : F] = q$ for distinct primes p and q, prove that $K \cap L = F$.
15. Let p and q be distinct primes. If $[F(\alpha) : F] = p$ and $[F(\beta) : F] = q$, prove that $[F(\alpha, \beta) : F] = pq$.

16. Prove that $[\mathbb{Q}(\sqrt{3}, \sqrt{5}) : \mathbb{Q}] = 4$.
17. Find an element $\alpha \in \mathbb{Q}(\sqrt{3}, \sqrt{5})$ for which $\mathbb{Q}(\sqrt{3}, \sqrt{5}) = \mathbb{Q}(\alpha)$. Then find $\min_{\mathbb{Q}}(\alpha)$.
18. Prove that every field homomorphism is one-to-one.
19. Prove Theorem 5.51.

References

1. Herstein I (1975) Topics in algebra, 2nd edn. Wiley, Hoboken
2. van der Waerden BL (1971) Modern algebra. Springer, Berlin

Chapter 6
Ruler and Compass Constructions

One remarkable application of abstract algebra arises in connection with four classical questions, originating with mathematicians of ancient Greece, about geometric constructions.

1. Is it possible to trisect an arbitrary constructible angle?
2. Is it possible to construct a square whose area is equal to that of a given constructible circle?
3. Which regular polygons are constructible?
4. Is it possible to construct a cube of exactly twice the volume of a given cube?

Questions 1–3 refer to constructions in the real plane $\mathbb{R}^2$ and Question 4 to constructions in $\mathbb{R}^3$. The tools we are allowed are only a compass and an uncalibrated ruler, i.e., devices for constructing circles and straight line segments, without measuring. The only constructions we can make come from a sequence of basic ones:

1. Connecting two given points with a straight line,
2. drawing the circle centered at a given point passing through another given point,
3. constructing a point as the intersection of two lines,
4. constructing a point (or points) as the intersection of a line and a circle, and
5. constructing a point (or points) as the intersection of two circles.

As an example of a construction, consider this familiar one:

Bisecting a line segment: Suppose that A and B are two distinct points in the plane. To bisect the line segment AB construct the circle centered at A passing through B and the one centered at B and passing through A. These circles intersect at two points and the line segment joining them bisects AB.

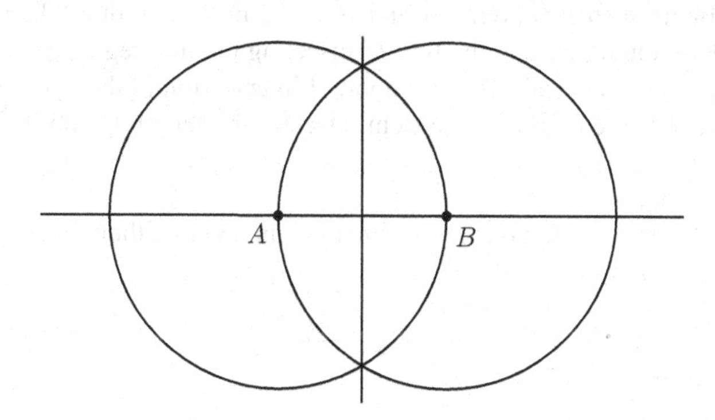

D.R. Finston and P.J. Morandi, *Abstract Algebra: Structure and Application*,
Springer Undergraduate Texts in Mathematics and Technology, DOI 10.1007/978-3-319-04498-9_6

Suppose that we identify two points A and B in the plane and declare that the distance between them is equal to 1. By drawing the line through A and B and using our unit distance AB as a reference, we can construct all points on this line whose distance from point A is an integer. Our first construction enables us to construct a sequence $\{A_i\}_{i=1}^{\infty}$ of points for which the distance from A to A_i is $(\frac{1}{2})^i$, and similar sequences for each of the new points we have constructed. In the exercises below you will find that it is possible to construct points separated by certain irrational distances as well.

Our investigation of constructibility proceeds by first using familiar construction techniques to establish a coordinate system on the plane, i.e., by constructing the points with integer coordinates. Then we examine the coordinates of point(s) lying on the intersection of two lines joining constructed points, on the intersection of a line with a circle of radius equal to a constructed distance centered at a constructed point, and on the intersection of two such circles. The connection with algebra comes from the fact that the collection of coordinates of all constructed points constitutes a field strictly between the fields of rational numbers and real numbers. The information we gain about this field will enable the investigation of the classical constructibility problems.

Exercises

1. Let A and B be points in the plane which we declare to be one unit apart. Denote by L_1 the line through A and B, and by $\mathbf{C}$ the circle centered at A passing through B (its radius is equal to 1). Let $B' \neq B$ be the other point of intersection of $\mathbf{C}$ with L_1 and L the perpendicular bisector of $B'B$. If D is the intersection of L and $\mathbf{C}$, show that the length of DB is equal to $\sqrt{2}$.
2. Complete the details of the following sketch of a proof of the irrationality of $\sqrt{2}$ alternative to the one in Exercise 3.3.7.

   (a) Prove that $0 < \sqrt{2} - 1 < 1$.
   (b) Prove that if b is a positive integer satisfying $b\sqrt{2} \in \mathbb{Z}$ then $b(\sqrt{2}-1)$ is also a positive integer satisfying $b(\sqrt{2}-1)\sqrt{2} \in \mathbb{Z}$.
   (c) Conclude that if b is a positive integer satisfying $b\sqrt{2} \in \mathbb{Z}$ then $b, b(\sqrt{2}-1), b(\sqrt{2}-1)^2, \ldots$ is an infinite decreasing sequence of positive integers.
   (d) Conclude that $\sqrt{2}$ is irrational.

6.1 Constructing a Coordinate System

Our first goal is to construct a coordinate system by choosing a length to be the unit length and to construct all of the points in the real plane $\mathbb{R}^2$ having integer coordinates relative to a fixed pair of perpendicular axes. Begin with two points A and B in the plane and draw the line L joining them. Declare L to be the horizontal axis and the line L' bisecting the line segment AB (constructed at the beginning of the chapter) the vertical axis. The point of intersection of the two axes is denoted O and referred to as the origin of the coordinate system. Declare the length of the line segment OA to be equal to 1.

The important construction:

Given a line L and a point A not on L, construct the unique line though A that is parallel to L:

A

L

applied to the horizontal and vertical axes L and L' enables the construction of our coordinate system. The details of the construction are in Exercises 1– 4.

Exercises

1. (Dropping a perpendicular) Construct two points P, Q on L by intersecting L with a circle centered at A. Bisect PQ as above and explain why the line joining A to the midpoint B of PQ is perpendicular to L.
2. Bisect QA and construct the line joining B and the midpoint of QA.
3. Intersect the circle centered at B whose radius is the length of QA with the line of the previous exercise. Now complete the construction of the line through A parallel to L, and verify the parallelism.
4. Explain how to construct all points (a, b) in $\mathbb{R}^2$ where a, b are integers.

Define a point (a, b) in the plane to be constructible if it can be reached from the integer grid points by a sequence of the basic constructions 1–5 above. Define a real number a to be constructible if $(a, 0)$ is a constructible point. Together with the first exercise of this chapter, which shows that $\sqrt{2}$ is a constructible number, and our bisection procedure, which implies that if a is constructible, then so is $a/2$, the grid construction shows that the constructible numbers form a set of real numbers, properly larger than the integers containing at least some noninteger rational numbers and some irrationals. Our next goal is to find an algebraic description of the set of constructible numbers.

6.2 The Field of Constructible Numbers

We have seen that ruler and compass constructions beginning with rational number lengths lead immediately to irrational lengths. That is, the coordinates of constructed points lie inside a proper field extension of $\mathbb{Q}$. The main results of this section and the next show that the set of these coordinates constitute a rather special subfield of $\mathbb{R}$.

Theorem 6.1. *The set C of constructible numbers is a field extension of the rational numbers.*

Because the set of constructible numbers contains the ring of integers, it suffices to show that C is closed under the field operations. We will show also that C is closed under another operation:

Theorem 6.2. *If c is a constructible number, then so is $\sqrt{c}$. In other words, C is closed under taking square roots.*

To begin the proof of the first theorem, note that if a is constructible, then so is $-a$: construct the lines joining $(a, 0)$ and to $(0, 1)$ and $(0, -1)$ and their parallels through $(0, -1)$ and $(0, 1)$, respectively. The latter intersect at $(-a, 0)$. Furthermore, if L is a line, P a constructible point on L, and a a constructible number, then the points on L that are at distance a from P are also constructible: using L as the horizontal axis and P as the origin of the coordinate system, the sequence of constructions producing $(\pm a, 0)$ yield the desired points. If b is also constructible, then using the horizontal axis as L and $(a, 0)$ as P, we see that $a + b$ is constructible, i.e., the constructible numbers are closed under addition.

Call a line constructible if it joins two constructible points and a circle constructible if its center is a constructible point and its radius a constructible number. Thus we can "mark off" any constructible length on a constructible line and construct a constructible circle centered at any constructible point.

Exercise 6.3. Given that A and B are constructible numbers, with $B \neq 0$, explain why the points and parallel line segments L_1 and L_2 in the diagram below are constructible.

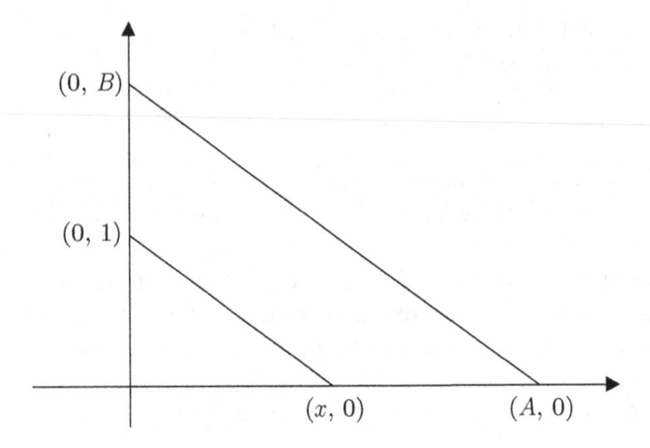

Exercise 6.4. Use facts about similar triangles to deduce the value of x and then explain why $\frac{A}{B}$ is constructible.

Since $AB = \frac{A}{\frac{1}{B}}$ we find that C is closed under multiplication as well, and hence have proved that the set of constructible numbers is a field extension of $\mathbb{Q}$.

Next we address the second theorem on closure under square roots.

Exercise 6.5. Suppose that a is a constructible number. Explain why the labeled points in the diagram below are constructible:

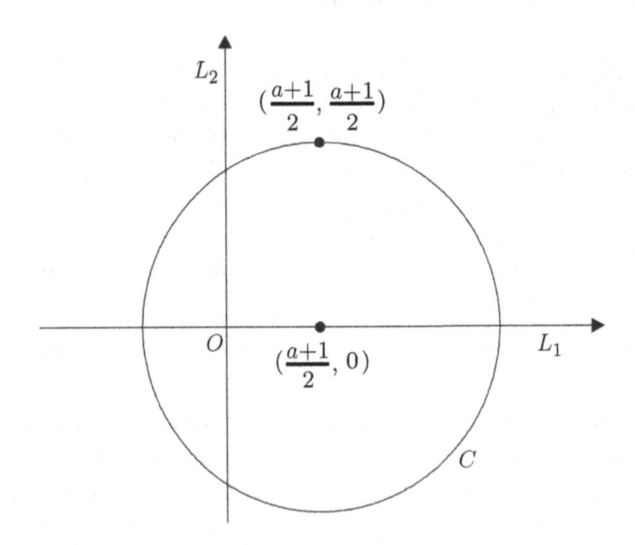

Exercise 6.6. From the remarks above, the circle **C** of radius $\frac{a+1}{2}$ centered at $(\frac{a+1}{2}, 0)$ is constructible. Write down its equation.

Exercise 6.7. Construct the intersection of the line through $(1, 0)$ that is parallel to L_2 with the circle **C**. Call this point $(1, y)$. Evaluate y to conclude that $\sqrt{a}$ is constructible and hence that C is closed under square roots.

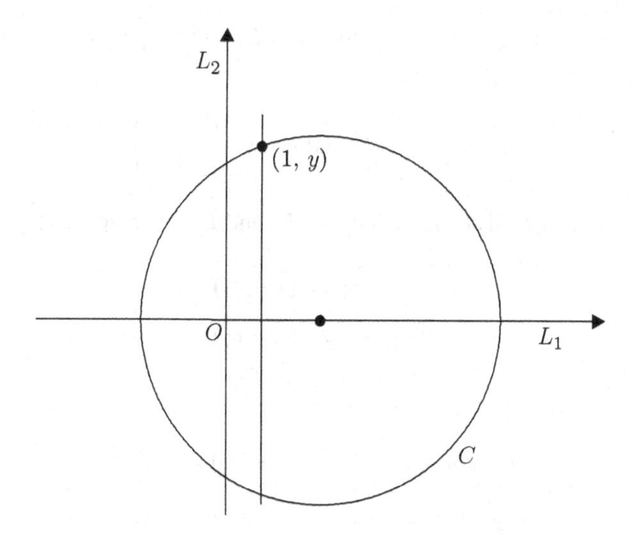

Example 6.8. $a = \frac{1+\sqrt{3}}{2+\sqrt{5}}$ is constructible.

Exercise 6.9. With a as in the example above, rationalize the denominator and calculate $[\mathbb{Q}(a) : \mathbb{Q}]$.

6.3 A Criterion for Constructibility

To address the classical construction problems we will derive a necessary criterion for a real number to be constructible. We can view the field C of constructible numbers as being built up in stages from the rational numbers. If a construction results in a point $P = (a, b)$, then we obtain the field extension $F = \mathbb{Q}(a, b)$ (which may be properly larger than $\mathbb{Q}$) all of whose elements are constructible. Moreover, any point in the plane $\mathbb{R}^2$ whose coordinates lie in F is constructible, since its coordinates are obtained by some sequence of field operations (addition, multiplication, subtraction, division) on constructible numbers, all of which result again in constructible numbers. We can make further constructions using these points to obtain larger field extensions. A field K so constructed will be called a *constructible field*, and the points of $\mathbb{R}^2$ whose coordinates lie in K, all of which are constructible, will be called *the plane of* K, denoted $\mathbb{P}(K)$. But how are the various constructible fields related?

To answer this question, let K be some field of constructible numbers. We determine the constructible numbers adjoined to K by the coordinates of points produced by the basic constructions applied to $\mathbb{P}(K)$:

1. The intersection of two nonparallel lines in $\mathbb{P}(K)$ (i.e., lines joining points whose coordinates lie in K),
2. the intersection of a line in $\mathbb{P}(K)$ with a circle in $\mathbb{P}(K)$ (i.e., a line as in 1 and a circle whose center is in $\mathbb{P}(K)$ and whose radius is in K), and
3. the intersection of two circles in $\mathbb{P}(K)$.

For points produced as in (1), suppose that the lines L_1 and L_2 pass through (a_1, b_1) and (c_1, d_1) and (a_2, b_2) and (c_2, d_2), respectively. Set

$$m_1 = \frac{(d_1 - b_1)}{(c_1 - a_1)} \text{ and } m_2 = \frac{(d_2 - b_2)}{(c_2 - a_2)},$$

noting that these slopes lie in K. Then equations for L_1 and L_2 are, respectively,

$$y - d_1 = m_1(x - c_1)$$
$$y - d_2 = m_2(x - c_2).$$

and they intersect where

$$d_1 + m_1(x - c_1) = d_2 + m_2(x - c_2).$$

Solving for x, we obtain

$$d_1 - d_2 + m_2c_2 - m_1c_1 = x(m_2 - m_1)$$
$$\frac{d_1 - d_2 + m_2c_2 - m_1c_1}{(m_2 - m_1)} = x.$$

Since the left-hand side involves field operations on elements of K, we find that x (and therefore also $y = d_1 + m_1(x - c_1)$) is also in K. Thus a point constructed as in (1) lies in $\mathbb{P}(K)$ and we don't enlarge the field K.

Coordinates of points produced by constructions (2) and (3) are investigated in the following exercises and lemma.

Exercise 6.10. Show that a line in $\mathbb{P}(K)$ has an equation of the form $ax + by + c = 0$ where a, b, c all lie in K.

Exercise 6.11. Show that a circle in $\mathbb{P}(K)$ has an equation of the form $x^2+y^2+dx+ey+f = 0$ where d, e, f all lie in K.

Exercise 6.12. Show that the coordinates of the points of intersection of a line L in $\mathbb{P}(K)$ with a circle $\mathbf{C}$ in $\mathbb{P}(K)$ solve a quadratic equation whose coefficients lie in K. First consider the case where the equation of L is $ax+by+c = 0$ with $a \neq 0$ and then the case where $a = 0$.

Exercise 6.13. Given that $\mathbf{C}_1$ and $\mathbf{C}_2$ are intersecting circles in $\mathbb{P}(K)$, suppose that $\mathbf{C}_1$ has equation

$$x^2 + y^2 + ax + by + c = 0$$

and $\mathbf{C}_2$ has equation

$$x^2 + y^2 + dx + ey + f = 0$$

where all of a, b, c, d, e, f lie in K.

1. If $a \neq d$ show that

(a) The x coordinates of the points of intersection have the form

$$x = \frac{(c - f) + (b - e)y}{d - a}$$

(b) The y coordinates of the points of intersection of $\mathbf{C}_1$ and $\mathbf{C}_2$ solve a quadratic equation whose coefficients lie in K.

2. If $a = d$, show that the y coordinates lie in K and that the x coordinates solve a quadratic equation whose coefficients lie in K.

The discussion concluding Sect. 5.4 shows that a polynomial with coefficients in the field K factors into linear factors in $K[x]$ if and only if all of its roots lie in K. Applying a simple case of this observation, the following lemma and theorem explain the special way in which constructible numbers arise.

Lemma 6.14. *Let F be a field extension of $\mathbb{Q}$ and $f = ax^2 + bx + c$ a quadratic polynomial with coefficients in F. If α is a root of f, then $[F(\alpha) : F] = 1$ if f is reducible and $[F(\alpha) : F] = 2$ if f is irreducible.*

Proof. See Exercise 4 of Sect. 5.4 for the case that f is reducible. Assume that f is irreducible and apply the quadratic formula to obtain the roots of f as

$$\alpha_1 = \frac{-b + \sqrt{b^2 - 4ac}}{2a},$$

$$\alpha_2 = \frac{-b - \sqrt{b^2 - 4ac}}{2a}.$$

Set $d = b^2 - 4ac$ and $\delta = \sqrt{d}$. Irreducibility of f implies that $\alpha_1, \alpha_2 \notin F$ and therefore $\delta \notin F$, since $F(\alpha_1) = F(\delta) = F(\alpha_2)$ as any field extension of F containing δ must also contain α_1 and vice-versa (similarly for α_2). From the fact that $\delta^2 \in F$, every element $\gamma \in F(\delta)$ can be written as $\gamma = \frac{r+s\delta}{u+v\delta}$ for certain r, s, u, v in F. By calculating $\left(\frac{r+s\delta}{u+v\delta}\right)\left(\frac{u-v\delta}{u-v\delta}\right)$, as one would "rationalize the denominator" over $\mathbb{Q}$, $\gamma = e + f\delta$ for some $e, f \in F$. Thus $1, \delta$ spans $F(\delta)$, but a dependence relation over F between 1 and δ (i.e., $e + f\delta = 0$ with $e, f \in F$ not both equal to 0) would imply $\delta \in F$. □

We have seen that given a constructible field K and its plane $\mathbb{P}(K)$, the coordinates of points that arise from a construction in $\mathbb{P}(K)$ solve linear or quadratic polynomials with coefficients in K. This gives us a numerical criterion that is necessarily satisfied by every constructible number, and can therefore be used to demonstrate that certain numbers are not constructible:

Theorem 6.15. *If α is a constructible number, then $[\mathbb{Q}(\alpha) : Q]$ is a power of 2.*

Proof. Since α is constructible a sequence of basic constructions will produce the point $(\alpha, 0)$. The constructions yield a sequence of field extensions

$$\mathbb{Q} = F_0 \subseteq F_1 \subseteq F_2 \subseteq \cdots \subseteq F_n$$

with $\alpha \in F_n$ and F_{i+1} obtained from F_i by solving a quadratic equation with coefficients in F_i. From the lemma we know that $[F(\alpha) : F] = 1$ or 2 so that repeated applications of the Dimension Formula yield $[F_n : \mathbb{Q}] = 2^m$ for some $m \leq n$. Since $\mathbb{Q} \subseteq \mathbb{Q}(\alpha) \subseteq F_n$ another application of the Dimension Formula shows that $[\mathbb{Q}(\alpha) : Q]$ must be a power of 2. □

Exercises

1. Use the Rational Roots Test to prove that $x^3 - 2 \in \mathbb{Q}[x]$ has no rational roots and is therefore irreducible over $\mathbb{Q}$.
2. Prove that $\sqrt[3]{2}$ is not a constructible number

6.4 Classical Construction Problems

The numerical criterion proved in Theorem 6.15 above enables us to solve all of the classical constructibility problems—the first three in the negative and the fourth through a complete description of the regular polygons which are constructible. Since the notion of constructible angle arises in these considerations we define it here.

Definition 6.16. An angle θ is said to be constructible if two lines can be constructed which intersect in that angle.

Exercise 6.17. Explain why the following construction shows that a $60°$ angle is constructible: Let $A = (0,0)$ and $B = (1,0)$. Denote by L the perpendicular bisector of AB and M the midpoint of AB. Construct the circle **C** centered at A passing through B, and let D be a point of intersection with L. Calculate the length of the line segment MD and show that the measure of angle MAD is $60°$.

6.4.1 Angle Trisection

While certain angles can be trisected (the trisection of a $180°$ angle was effectively accomplished in Exercise 6.17 above) there is no sequence of constructions that can be applied to trisect an arbitrary angle. To see this, we prove the following result.

Theorem 6.18. *An angle of* $60°$ *cannot be trisected by ruler and compass constructions.*

Proof. If it were possible to trisect a $60°$, then it would be possible to construct the right triangle OPQ below:

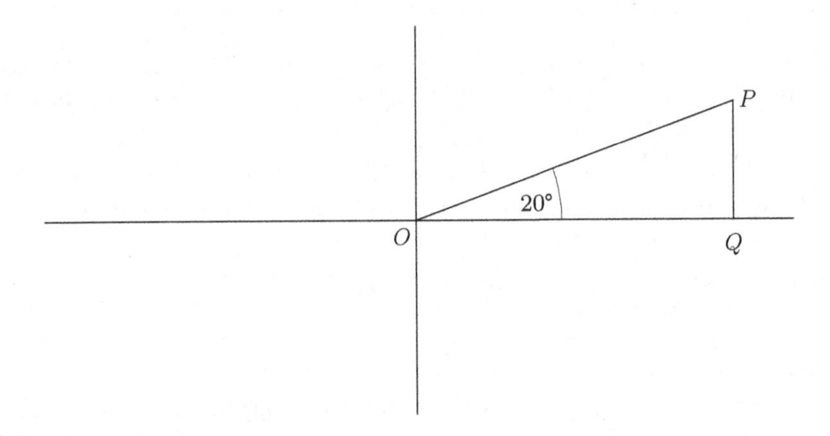

Since the coordinates of P are $(\cos 20°, \sin 20°)$, these numbers would also be constructible. An exercise in trigonometry using multiple angle formulas reveals that $\alpha = \cos 20°$ is a root of the cubic polynomial $f = 8x^3 - 6x - 1$. Indeed, for any angle x,

$$\begin{aligned}\cos 3x &= \cos(x+2x) = \cos x \cos 2x - \sin x \sin 2x \\ &= \cos x(\cos^2 x - \sin^2 x) - 2\sin^2 x \cos x \\ &= \cos^3 x - 3\sin^2 x \cos x \\ &= \cos^3 x - 3\cos x(1-\cos^2 x) \\ &= 4\cos^3 x - 3\cos x.\end{aligned}$$

We obtain for $x = 20°$ that

$$\frac{1}{2} = 4\alpha^3 - 3\alpha$$

from which it is clear α is a root of f.

Using the Rational Roots Test, one can show that f has no rational roots and is therefore irreducible as a polynomial with rational coefficients. Thus $\min_{\mathbb{Q}}(\alpha) = \frac{1}{8}f = x^3 - \frac{3}{4}x - \frac{1}{8}$ which has degree equal to 3. Corollary 5.46 yields $[\mathbb{Q}(\alpha) : \mathbb{Q}] = 3$ so that by our numerical criterion, α, and therefore the $20°$ angle, are not constructible. □

6.4.2 Duplicating a Cube

The problem asks whether, given a constructible cube, it is possible to construct a cube of precisely twice the volume of the given one. To effect such constructions it is necessary of course to extend our constructions from the plane to three-dimensional space. This is not a serious issue though. The problem is solved in the negative by demonstrating the following result.

Theorem 6.19. *The unit cube cannot be duplicated, i.e., a cube whose volume is equal to 2 cannot be constructed with ruler and compass.*

Proof. If it were possible to construct a cube whose volume is equal to 2, then each side of the cube would have length $\sqrt[3]{2}$. From Example 5.47 we know that.$[\mathbb{Q}(\sqrt[3]{2}) : \mathbb{Q}] = 3$, so that $\sqrt[3]{2}$ is not constructible, and hence that this construction is not possible. □

6.4.3 Squaring the Circle

At issue is whether, given a constructible circle, it is possible to construct a square with precisely the same area as the given circle. This problem is also resolved in the negative by considering the unit circle, whose area is π. A square of area equal to π would have a side of length $\sqrt{\pi}$ so that constructibility of said square is equivalent to constructibility of $\sqrt{\pi}$. Since the constructible numbers are closed under squaring and square roots, the issue rests on the constructibility of π. A famous but rather difficult theorem asserts that π is a transcendental number, i.e., π is not algebraic over $\mathbb{Q}$ so is not a root of any nonzero polynomial with rational coefficients. From Theorem 5.51 we know that the infinitely many powers $1, \pi, \pi^2, \ldots$ are linearly independent over $\mathbb{Q}$ (i.e., no finite collection of distinct powers is linearly dependent). Thus $[\mathbb{Q}(\pi) : \mathbb{Q}]$ is infinite (and not a power of 2).

6.4.4 Constructible Polygons

Consider a regular n-gon $\mathcal{P}_n$ (i.e., $\mathcal{P}_n$ is a polygon with n sides of equal length) inscribed within the unit circle with one vertex at the point $(1, 0)$.

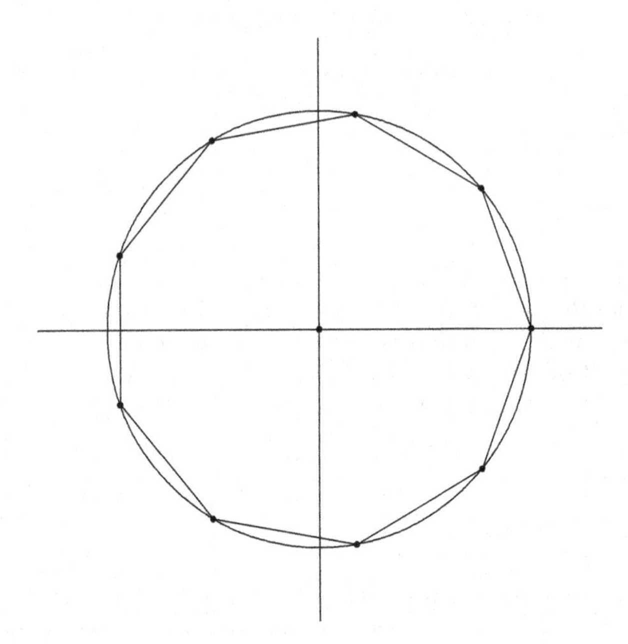

It is most convenient here to use radian measure for angles so that the set of vertices of $\mathcal{P}_n$ is exactly $\{(\cos \frac{2\pi j}{n}, \sin \frac{2\pi j}{n}) : j = 0, \ldots n-1\}$. Thus the regular n-gon is constructible if and only if $\cos \frac{2\pi j}{n}$ and $\sin \frac{2\pi j}{n}$ are constructible numbers. With i denoting the complex square root of -1 and θ any real number, de Moivre's theorem gives us $e^{i\theta} = \cos\theta + i \sin\theta$. In particular, $(e^{\frac{2\pi i}{n}})^j = \cos \frac{2\pi j}{n} + i \sin \frac{2\pi j}{n}$ so that the coordinates of the vertices of $\mathcal{P}_n$ all lie in the field $F_n = \mathbb{Q}(\cos \frac{2\pi}{n}, \sin \frac{2\pi}{n})$, a subfield of $\mathbb{R}$. Let K_n be the field $\mathbb{Q}(e^{\frac{2\pi i}{n}}, i)$. Since for every θ,

$$\begin{aligned}(e^{i\theta})^{-1} &= e^{-i\theta} \\ &= \cos\theta - i \sin\theta,\end{aligned}$$

we see that $\cos \frac{2\pi}{n} - i \sin \frac{2\pi}{n}$ also lies in K_n. From $\cos\theta = \frac{e^{i\theta}+e^{-i\theta}}{2}$ and $\sin\theta = \frac{e^{i\theta}-e^{-i\theta}}{2i}$ it follows that the coordinates of all of the vertices of $\mathcal{P}_n$ lie in K_n and also that $F_n(i) = K_n$

In order to investigate the conditions under which $[F_n : \mathbb{Q}]$ is a power of 2, first observe that $[K_n : F_n] = 2$ because F_n is a subfield of the real numbers and $1, i$ constitute a basis for K_n over F_n. Similarly, if i is not in the field $\mathbb{Q}(e^{\frac{2\pi i}{n}})$, then also

$$[\mathbb{Q}(e^{\frac{2\pi i}{n}}, i) : \mathbb{Q}(e^{\frac{2\pi i}{n}})] = 2.$$

As a consequence we see that

$$\begin{aligned}[F_n : \mathbb{Q}] &= \frac{[K_n : \mathbb{Q}]}{2} \\ &= \frac{[\mathbb{Q}(e^{\frac{2\pi i}{n}}, i) : \mathbb{Q}]}{2} \\ &= [\mathbb{Q}(e^{\frac{2\pi i}{n}}) : \mathbb{Q}] \text{ or } \frac{[\mathbb{Q}(e^{\frac{2\pi i}{n}}) : \mathbb{Q}]}{2},\end{aligned}$$

depending on whether or not i is in the field $\mathbb{Q}(e^{\frac{2\pi i}{n}})$. Recalling that $F_n = \mathbb{Q}(\cos\frac{2\pi}{n}, \sin\frac{2\pi}{n})$, we have proved the following proposition.

Proposition 6.20. $[\mathbb{Q}(\cos\frac{2\pi}{n}, \sin\frac{2\pi}{n}) : \mathbb{Q}]$ *is a power of 2 if and only if* $[\mathbb{Q}(e^{\frac{2\pi i}{n}}) : \mathbb{Q}]$ *is a power of 2. In particular, if the regular n-gon is constructible, then* $[\mathbb{Q}(\cos\frac{2\pi}{n}, \sin\frac{2\pi}{n}) : \mathbb{Q}]$ *is a power of 2.*

The field extensions $\mathbb{Q}(e^{\frac{2\pi i}{n}})$ of $\mathbb{Q}$, known as the cyclotomic extensions, are very well studied for their applications in number theory and other branches of mathematics. The integer $[\mathbb{Q}(e^{\frac{2\pi i}{n}}) : \mathbb{Q}]$, which is denoted by $\varphi(n)$, has great significance in number theory, some of which is investigated in later chapters. Remarkably, $\varphi(n)$ is equal to the number of positive integers less than n whose greatest common divisor with n is equal to 1. So, for instance, $\varphi(3) = \varphi(4) = \varphi(6) = 2$, $\varphi(5) = \varphi(8) = 4$, $\varphi(7) = \varphi(9) = 6$. In general, if n is prime, then $\varphi(n) = n - 1$.

The contrapositive of the proposition gives us a method to show that certain n-gons are not constructible.

Example 6.21. The regular 9-gon is not constructible. To see this let $\alpha = \cos\frac{2\pi}{9}$. Note that

$$-\frac{1}{2} = \cos\frac{2\pi}{3} = \cos 3(\frac{2\pi}{9}) = 4\alpha^3 - 3\alpha$$

so that α is a root of $8x^3 - 6x + 1$. Once again the Rational Roots Test shows that this polynomial is irreducible and arguments similar to those used in the trisection problem show that α is a root of no quadratic polynomial with rational coefficients. We again obtain that $[\mathbb{Q}(\alpha) : \mathbb{Q}]$ is equal to three so that

$$[\mathbb{Q}(\cos\frac{2\pi}{9}, \sin\frac{2\pi}{9}) : \mathbb{Q}] = [\mathbb{Q}(\cos\frac{2\pi}{9}, \sin\frac{2\pi}{9}) : \mathbb{Q}(\cos\frac{2\pi}{9})][\mathbb{Q}(\cos\frac{2\pi}{9}) : \mathbb{Q}]$$

is not a power of 2.

Exercise 6.22. Use the Pythagorean theorem to show that $[\mathbb{Q}(\cos\theta, \sin\theta) : \mathbb{Q}(\cos\theta)] \leq 2$. Conclude that it suffices to consider $[\mathbb{Q}(\cos\frac{2\pi}{n}) : \mathbb{Q}]$ in Proposition 6.20.

While Proposition 6.20 gives the necessary condition that $\varphi(n)$ must be a power of 2 for the regular n-gon to be constructible, the converse is also true:

Theorem 6.23. $\mathbb{Q}(e^{\frac{2\pi i}{n}})$ *is a constructible field, and therefore the regular n-gon is constructible, if and only if* $\varphi(n)$ *is a power of 2.*

A complete proof of this theorem is beyond the scope of this book, but using it we see that the 7-gon is not constructible, while the 17-gon is.

Exercises

1. Prove that the angle θ is constructible if and only if $\cos\theta$ (and $\sin\theta$) are constructible numbers.
2. Prove that a 90° angle can be trisected.
3. Prove that a 45° angle can be trisected.

References

1. Hobson EW (1913) Squaring the circle. Cambridge University Press, Cambridge
2. Klein F (1895) Ausgewählte Fragen der Elementargeometrie. Göttingen
3. Niven I (1939) The transcendence of *pi*. Am Math Monthly 46(8):469–471
4. Niven I (1947) The irrationality of *pi*. Bull Am Math Soc 53(6):509
5. Richmond HW (1893) A construction for a regular polygon of seventeen sides. Q J Pure Appl Math 26:206–207

Chapter 7
Cyclic Codes

In this chapter we will build codes from quotient rings of $\mathbb{Z}_2[x]$. One advantage of this construction will be that we can guarantee a certain degree of error correction. We first make a connection between words and elements of such a quotient ring.

7.1 Introduction to Cyclic Codes

Let n be a positive integer. If $f(x) \in \mathbb{Z}_2[x]$ is a polynomial of degree n, then by the Division Algorithm, each element of the quotient ring $\mathbb{Z}_2[x]/(f(x))$ can be represented uniquely as the coset of a polynomial of degree less than n. In other words, if $g(x) \in \mathbb{Z}_2[x]$, then we may write $g(x) = q(x)f(x) + r(x)$ with $\deg(r(x)) < \deg(f(x)) = n$. Then $g(x) + (f(x)) = r(x) + (f(x))$; this yields the desired representation for the coset $g(x) + (f(x))$. Uniqueness follows from the Division Algorithm. However, we recall the idea. If $r(x) + (f(x)) = r'(x) + (f(x))$ with $\deg(r), \deg(r') < n$, then $r(x) + r'(x) \in (f(x))$. Thus, $f(x)$ divides $r(x) + r'(x)$. However, this forces $\deg(r + r') \geq n$ unless $r + r' = 0$. Since $\deg(r + r') \not\geq n$, we must conclude that $r + r' = 0$, so $r' = r$. Therefore, there is a 1-1 correspondence between elements of $\mathbb{Z}_2[x]/(f(x))$ and polynomials of degree $< n$. On the other hand, if $\mathbb{Z}_2[x]_n$ is the set of polynomials of degree $< n$, then identifying a polynomial with its n-tuple of coefficients provides a 1-1 correspondence between $\mathbb{Z}_2^n$ and $\mathbb{Z}_2[x]_n$ given by

$$(a_0, \ldots, a_{n-1}) \mapsto a_0 + a_1 x + \cdots + a_{n-1} x^{n-1}$$

We will use the polynomial $f(x) = x^n + 1$ in our discussion below and then identify $\mathbb{Z}_2[x]/(x^n + 1)$ with the set of words of length n. Since words in natural language are not typically written as vectors, the notation $a_1 \cdots a_n$ will be used in place of $(a_1, \ldots, a_n)$ for vectors in $\mathbb{Z}_2^n$ viewed as words in a linear code. For instance, in $\mathbb{Z}_2[x]/(x^7 + 1)$ the word 101001 is identified with the residue class of $1 + x^2 + x^5$.

Electronic supplementary material The online version of this chapter (doi:10.1007/978-3-319-04498-9_7) contains supplementary material, which is available to authorized users. The supplementary material can also be downloaded from http://extras.springer.com.

D.R. Finston and P.J. Morandi, *Abstract Algebra: Structure and Application*,
Springer Undergraduate Texts in Mathematics and Technology, DOI 10.1007/978-3-319-04498-9_7

Define a map $\pi : \mathbb{Z}_2^n \to \mathbb{Z}_2^n$ by $\pi(a_1 \cdots a_n) = a_n a_1 \cdots a_{n-1}$. This is the (right) shift function because it shifts the components of a vector to the right, and moves the last component to the front. In terms of the representation of $\mathbb{Z}_2^n$ as $\mathbb{Z}_2[x]/(x^n + 1)$, the map π is given by

$$\pi(a_0 + \cdots + a_{n-1}x^{n-1} + (x^n + 1)) = a_{n-1} + a_0 x + \cdots + a_{n-2}x^{n-1} + (x^n + 1).$$

Definition 7.1. Let C be a linear code of length n. Then C is a cyclic code if for every $v \in C$, we have $\pi(v) \in C$.

Cyclic codes with good error correcting and decoding properties arise naturally from quotients of polynomial rings.

Example 7.2. The codes $\{000, 101, 110, 011\}$ and $\{00000, 11111\}$ are cyclic codes.

Example 7.3. The Hamming code is the linear code

$$C = \{0111100, 0100101, 0011001, 0010110, 0001111, 0110011, 1010101, 1111111,$$
$$1100110, 1011010, 1000011, 0101010, 1001100, 1110000, 0000000, 1101001\}.$$

Since $0111100 \in C$ but $\pi(0111100) = 0011110 \notin C$ w see that C is not cyclic.

Example 7.4. Recall that the Hamming code is the nullspace of

$$H = \begin{pmatrix} 0\,0\,0\,1\,1\,1\,1 \\ 0\,1\,1\,0\,0\,1\,1 \\ 1\,0\,1\,0\,1\,0\,1 \end{pmatrix}.$$

Rearranging the columns of H permutes the entries of its nullspace vectors and can result in a cyclic code. Let

$$A = \begin{pmatrix} 1\,0\,0\,1\,0\,1\,1 \\ 0\,1\,0\,1\,1\,1\,0 \\ 0\,0\,1\,0\,1\,1\,1 \end{pmatrix}.$$

Then the nullspace of A is

$$C = \{1010001, 0110100, 1101000, 0010111, 1000110, 0100011, 1111111, 0001101,$$
$$1001011, 1100101, 1110010, 0111001, 1011100, 0101110, 0011010, 0000000\}.$$

A calculation will show that this code is cyclic. One way to verify this is to note that if v is the first vector listed, then $\pi(v), \pi^2(v), \ldots, \pi^6(v)$ are all in the code. Note that $\pi^7(v) = v$. Also, if w is the fourth vector, then $\pi(w), \ldots, \pi^6(w), \pi^7(w) = w$ are also in the code, and are all different from the 7 vectors $v, \ldots, \pi^6(v)$. Finally, the two remaining vectors are 0000000 and 1111111, which are invariant under π. The description

$$C = \{\pi^i(v) : 0 \le i \le 6\} \cup \{\pi^i(w) : 0 \le i \le 6\} \cup \{0000000, 1111111\}.$$

together with a short proof shows that C is cyclic. Admittedly, this ad-hoc demonstration that the modified Hamming code is cyclic is not very natural. A closer examination of the connection between quotients of polynomial rings and cyclic codes will clarify this issue.

The utility of the polynomial $x^n + 1$ emerges in the next proposition. The proof, which is left to the reader, is a straightforward comparison of $\pi(f(x) + (x^n + 1))$ with $xf(x) + (x^n + 1)$ using the identification of $\mathbb{Z}_2^n$ with $\mathbb{Z}_2[x]/(x^n + 1)$. Set $I = (x^n + 1)$ and recall that in $\mathbb{Z}_2[x]/I$ we have $x^n = -1 = 1$.

Proposition 7.5. *Under the identification of* $\mathbb{Z}_2^n = \mathbb{Z}_2[x]/I$*, the map* π *is given by* $\pi(f(x) + I) = xf(x) + I$.

Theorem 7.6. *Let* C *be a cyclic code of length* n*. Then* C *is an ideal of* $\mathbb{Z}_2[x]/I$*. Conversely, every ideal of* $\mathbb{Z}_2[x]/I$ *is a cyclic code.*

Proof. Since C is closed under addition, to show that C is an ideal it suffices to show that for $f + I \in C$ and $g + I \in \mathbb{Z}_2[x]/I$, $gf + I = (g + I)(f + I) \in C$. Writing $g = \sum_{i=0}^{n-1} a_i x^i$, and using the linearity of C, it suffices to show that $x^i f + I \in C$ for each $0 \le i \le n - 1$. By the proposition, $x^i f + I = \pi^i(f + I)$ which lies in C by our assumption that C is cyclic.

Conversely, if C is an ideal of $\mathbb{Z}_2[x]/I$, then C is a $\mathbb{Z}_2$ vector subspace of $\mathbb{Z}_2[x]/I$ hence a linear code. But closure under multiplication by $x + I$ and the proposition shows that C is also cyclic. □

The following theorem and its corollaries are instances of the "isomorphism theorems" important throughout modern algebra.

Theorem 7.7. *Let* R *be a commutative ring and* I *an ideal of* R*. Then the ideals of* R/I *are in one-to-one correspondence with the ideals of* R *which contain* I*. The correspondence is given as follows: For* J *an ideal of* R *containing* I*, the corresponding ideal of* R/I *is* $J/I = \{a + I : a \in J\}$*. For* $\overline{J}$ *an ideal of* R/I*, the corresponding ideal of* R *is* $\{a \in R : a + I \in \overline{J}\}$.

Remark 7.8. Three things must be shown:

1. Given J an ideal of R containing I, J/I is an ideal of R/I.
2. Given $\overline{J}$ an ideal of R/I, $J = \{a \in R : a + I \in \overline{J}\}$ is an ideal of R containing I.
3. The processes of 1. and 2. are inverse to each other.

Proof. For 1., observe that for $a, b \in J$, $(a + I) + (b + I) = (a + b) + I \in J/I$ because J is closed under addition. Similarly, for $c + I \in R/I$, $(c + I)(a + I) = ca + I \in J/I$.

For 2., suppose that $a, b \in J$ (i.e., $a + I, b + I \in \overline{J}$), and that $c \in R$. Then $(a + b) + I = (a + I) + (b + I) \in \overline{J}$ so that $a + b \in J$. Also, $ca + I = (c + I)(a + I) \in \overline{J}$ so that $ca \in J$. It is clear that J contains I, since $a + I = 0 + I$ lies in J bar for every a in I.

To establish 3., note first that given an ideal J of R, J is certainly contained in $\{a \in R : a + I \in J/I\}$. On the other hand, this set is equal to $\{a \in R : a + I = b + I$ for some $b \in J\}$. But $a + I = b + I$ implies that $a - b \in I \subseteq J$ and, since $b \in J$, we obtain that $a \in J$ as well. Thus $J = \{a \in R : a + I \in J/I\}$.

For an ideal $\overline{J}$ of R/I and $J = \{a \in R : a + I \in \overline{J}\}$ the corresponding ideal of R, we show that $\overline{J} = J/I$. Observe that $a + I \in \overline{J}$ if and only if $a \in J$, and this holds if and only if $a + I \in J/I$ by definition of this ideal. □

We have seen that for F a field, all ideals of $F[x]$ are principal. The same holds for quotient rings of $F[x]$.

Corollary 7.9. *Let* F *be a field,* I *an ideal of* $F[x]$*, and* $\overline{J}$ *an ideal of* $F[x]/I$*. Then* $\overline{J}$ *is a principal ideal.*

Proof. By Theorem 7.7, $\overline{J} = J/I$ for some ideal J of $F[x]$. Since every ideal of $F[x]$ is principal $J = (g)$ for some $g \in F[x]$. If $a + I \in J/I$, then $a \in J$ so that $a = gh$ for some $h \in F[x]$. Thus $a + I = gh + I = (g + I)(h + I)$, which lies in the principal ideal $(g + I)$ of $F[x]/I$. Since $g \in J$, we obtain immediately that $(g + I) \in J/I$. □

Corollary 7.10. *Let C be a cyclic code of length n. Then there is a divisor g of $f(x) = x^n + 1$ in $\mathbb{Z}_2[x]$ so that $C = (g)/(f)$. Conversely every divisor of f gives rise to a cyclic code in this manner.*

Proof. Since C is an ideal of $\mathbb{Z}_2[x]/(f)$ it is principal, generated say by g, and $(f) \subset (g)$. But $(f) \subset (g)$ means precisely that g divides f. □

Definition 7.11. The generator polynomial of a cyclic code C of length n is a generator of C as an ideal of $\mathbb{Z}_2[x]/(x^n + 1)$.

Proposition 7.12. *Let C be a cyclic code of length n with generator polynomial g. Then $[g(x), xg(x), \ldots, x^{n-1-\deg(g)}g(x)]$ is a basis for C; therefore,* $\dim(C) = n - \deg(g)$.

7.2 Finite Fields

Recall Proposition 5.24, which says that if $p(x)$ is an irreducible polynomial in $\mathbb{Z}_2[x]$, then the quotient ring $\mathbb{Z}_2[x]/(p(x))$ is a field. Recall that the degree of $\mathbb{Z}_2[x]/(p(x))$ as a field extension of $\mathbb{Z}_2$ (i.e., its dimension as a $\mathbb{Z}_2$ vector space) is equal to the degree of $p(x)$. Therefore, if $\deg(p(x)) = n$, there are 2^n elements in this field. We will use this idea to construct finite fields of size a power of 2 as a tool in building codes. We remark that if $\mathbb{Z}_2[x]/I$ is such a field, then for any element $f(x) + I$, then $(f(x) + I) + (f(x) + I) = (f(x) + f(x)) + I$. However, since $f(x) \in \mathbb{Z}_2[x]$, we have $f(x) + f(x) = 0$. Thus, any element α of $\mathbb{Z}_2[x]/I$ satisfies $\alpha + \alpha = 0$, or $-\alpha = \alpha$. This fact will be used frequently.

We will denote by GF(q) a finite field with q elements ($q = 2^n$ where $n = [\mathrm{GF}(q) : \mathbb{Z}_2]$). To produce such fields requires the production of irreducible polynomials of degree n in $\mathbb{Z}_2[x]$. Given a polynomial $p(x) \in \mathbb{Z}_2[x]$, we can determine if $p(x)$ is irreducible by using a computer algebra package or by a tedious calculation if $\deg(p)$ is not too large. However, if $\deg(p) \leq 3$, we have an easy test for irreducibility.

Proposition 7.13. *Let F be a field, and let $p(x) \in F[x]$ with $2 \leq \deg(p) \leq 3$. Then $p(x)$ is irreducible over F if and only if $p(x)$ has no roots in F.*

Proof. One direction is easy. If $a \in F$ with $p(a) = 0$, then $x - a$ divides $p(x)$, and so $p(x) = (x - a)q(x)$ for some polynomial $q(x)$. The assumption on the degree of $p(x)$ shows that $q(x)$ is not a constant polynomial. Therefore, $p(x)$ is reducible. Conversely, suppose that $p(x)$ is reducible. Then we may factor $p(x) = f(x)g(x)$ with each of f and g nonconstant. Since $\deg(p) \leq 3$ and $\deg(p) = \deg(f) + \deg(g)$, we conclude that either $\deg(f) = 1$ or $\deg(g) = 1$. Suppose that $\deg(f) = 1$. Then $f(x) = ax + b$ for some $a, b \in F$ with $a \neq 0$. Then $f(x)$ has a root $-ba^{-1} \in F$; this element is then also a root of $p(x)$. A similar conclusion holds if $\deg(g) = 1$. □

Example 7.14. The polynomial $x^2 + x + 1$ is irreducible over $\mathbb{Z}_2$; thus, we obtain the field $\mathbb{Z}_2[x]/(x^2 + x + 1)$. Its elements are the cosets $0 + I, 1 + I, x + I, x + 1 + I$. We write the first two cosets as 0 and 1 and the latter two as a and b. We then see that $+$ and $\cdot$ are given by the tables below.

$+$	0	1	a	b
0	0	1	a	b
1	1	0	b	a
a	a	b	0	1
b	b	a	1	0

$\cdot$	0	1	a	b
0	0	0	0	0
1	0	1	a	b
a	0	a	b	1
b	0	b	1	a

These tables are exactly those of Example 3.30. Those tables were built from this quotient ring.

One important property of finite fields is the existence of primitive elements.

Definition 7.15. A primitive element of a finite field F is an element α so that every nonzero element of F is a power of α.

For convenience, we write F^* for the set of nonzero elements of F. Referring to Example 7.14, both a and b are primitive elements, since $F^* = \{1, a, b\}$ and $b = a^2$ and $a = b^2$. Note that $1 = a^0 = b^0$, so 1 is always a power of any nonzero element.

Example 7.16. The polynomial $x^3 + x + 1$ is irreducible over $\mathbb{Z}_2$ since it does not have a root in $\mathbb{Z}_2$. Thus, if $I = (x^3 + x + 1)$, we may form the field $\mathbb{Z}_2[x]/I$. This field has 8 elements. One representation of these elements is as the cosets of polynomials of degree less than 3. Alternatively, if $\alpha = x + I$, then we have the following table.

Power of α	Coset representative
α^0	1
α^1	x
α^2	x^2
α^3	$x + 1$
α^4	$x^2 + x$
α^5	$x^2 + x + 1$
α^6	$x^2 + 1$
α^7	1

To help see how to make these calculations, we point out that $\alpha^3 = (x+I)^3 = x^3+I$. However, $x^3 + x + 1 \in I$; therefore, $x^3 + I = x + 1 + I$. We can use this to calculate further powers of α. For example,

$$\begin{aligned}\alpha^4 &= \alpha^3 \cdot \alpha \\ &= (x + 1 + I)(x + I) \\ &= (x + 1)x + I \\ &= x^2 + x + I.\end{aligned}$$

By using this idea, along with the relation $x^3 + I = x + 1 + I$, we can obtain all the powers of α. Alternatively, we can use the Division Algorithm. For instance, since $\alpha^5 = x^5 + I$, we calculate that $x^5 = (x^3 + x + 1)(x^2 + 1) + x^2 + x + 1$, yielding $x^5 + I = x^2 + x + 1 + I$.

It is true that every finite field has a primitive element. However, the proof of this fact involves more complicated ideas than we will consider, so we do not give the proof. However, for any example, we will be able to produce a primitive element, either by a hand calculation or by using a computer algebra package.

Lemma 7.17. *Let α be a primitive element of* GF(2^n). *Then $\alpha^{2^n-1} = 1$. Furthermore, $\alpha^r \neq 1$ for all r with $1 \leq r < 2^n - 1$.*

Proof. Since GF$(2^n)^* = \{\alpha^i : i \geq 1\}$ is a finite set, there are positive integers $r < s$ with $\alpha^r = \alpha^s$. Then $\alpha^{s-r} = 1$. Thus, by the well-ordering property of $\mathbb{Z}$, there is a smallest integer t satisfying $\alpha^t = 1$. We next show that GF$(2^n)^* = \{\alpha^i : 0 \leq i < t\}$. To see this, if $i \in \mathbb{Z}$, then by the Division Algorithm, there are integers q and r with $i = qt + r$ and $0 \leq r < t$. Then $\alpha^i = \alpha^{qt+r} = (\alpha^t)^q \alpha^r = \alpha^r$ since $\alpha^t = 1$. This proves the claim. Furthermore, if $0 \leq i < j < t$, then $\alpha^i \neq \alpha^j$, since if $\alpha^i = \alpha^j$, then $\alpha^{j-i} = 1$, contradicting minimality of t. This proves that $\{\alpha^i : 0 \leq i < t\}$ has exactly t elements. Since this set is equal to GF$(2^n)^*$, which has $2^n - 1$ elements, we see that $t = 2^n - 1$. This proves that $\alpha^{2^n-1} = \alpha^t = 1$. Furthermore, we have shown that if $0 < i < t$, then $\alpha^0 \neq \alpha^i$, showing that $\alpha^i \neq 1$ if $0 \leq i < t = 2^n - 1$. □

7.3 Minimal Polynomials and Roots of Polynomials

Let I be the ideal $(x^2 + x + 1)$ in $\mathbb{Z}_2[x]$ and F the field $\mathbb{Z}_2[x]/I$ described in the previous section. If α is the coset of x, then we note that

$$\begin{aligned}\alpha^2 + \alpha + 1 &= (x^2 + I) + (x + I) + (1 + I) = x^2 + x + 1 + I \\ &= 0 + I.\end{aligned}$$

Therefore, α is a root of the polynomial $x^2 + x + 1$. In particular, α is algebraic over $\mathbb{Z}_2$. In fact, we note that each of the three nonzero elements of F is a root of $x^3 + 1 = (x + 1)(x^2 + x + 1)$. Moreover, 0 is the root of the polynomial x. As we now see, each element of a finite field is algebraic over $\mathbb{Z}_2$, hence a root of a nontrivial polynomial in $\mathbb{Z}_2[x]$. Since the base field is understood to be Z_2, the minimal polynomial min $_{Z_2}(\alpha)$ of an element alpha of an algebraic extension of Z_2 will be denoted $m_\alpha(x)$.

Theorem 7.18. *Let $\alpha \in$ GF(2^n). Then α is algebraic over $\mathbb{Z}_2$ and the minimal polynomial $m_\alpha(x) := \min_{\mathbb{Z}_2}(\alpha)$ divides $x^{2^n} + x$.*

Proof. If $\alpha = 0$, then $m_\alpha(x) = x$, and this clearly divides $x^{2^n} + x$. So, suppose that $\alpha \neq 0$. We give a proof assuming the existence of a primitive element for every finite field. Suppose that β is a primitive element of F. Then $\beta^{2^n-1} = 1$ by Lemma 7.17. Since $\alpha \neq 0$, there is an i with $\alpha = \beta^i$. Thus,

$$\alpha^{2^n-1} = (\beta^i)^{2^n-1} = (\beta^{2^n-1})^i = 1^i = 1.$$

This proves that α is a root of $x^{2^n-1} + 1$. Multiplying by x yields $x^{2^n} + x$, and α is then also a root of this polynomial. Since by definition $m_\alpha(x)$ generates the ideal of all polynomials in $\mathbb{Z}_2$ having α as a root, we see that $m_\alpha(x)$ divides $x^{2^n} + x$. □

We give an alternative proof of the previous theorem without using the existence of primitive elements, partly to be more complete, and also because it is a look ahead to group theory. The existence of a primitive element was used only to prove that $\alpha^{2^n-1} = 1$. We verify this fact without using primitive elements.

Proposition 7.19. *Let F be a finite field with $|F^*| = q$. Then $\alpha^q = 1$ for each $\alpha \in F^*$.*

Proof. List the elements of $F^* = \{a_1, \ldots, a_q\}$ and let α be any element of this set. We claim that $F^* = \{\alpha a_1, \ldots, \alpha a_q\}$. To see why, we clearly have $\{\alpha a_1, \ldots, \alpha a_q\} \subseteq F^*$ since $\alpha \neq 0$. For the reverse inclusion, we note that, given i, the element $a_i \alpha^{-1} \in F^*$, so there is a j with $a_i \alpha^{-1} = a_j$. Then $a_i = \alpha a_j$. This proves the claim. Next, since multiplication in F^* is commutative (and associative), multiplying all of the element of F^* together, we see that

$$a_1 \cdots a_q = (\alpha a_1) \cdots (\alpha a_q) = \alpha^q (a_1 \cdots a_q).$$

Cancellation yields $\alpha^q = 1$, as desired. □

Example 7.20. Consider the field $\mathrm{GF}(8) = \mathbb{Z}_2[x]/(x^3 + x + 1) = \mathbb{Z}_2[x]/I$. As we saw in Example 7.16, $\alpha = x + I$ is a primitive element. By Theorem 7.18, the minimal polynomial of each element of GF(8) divides $x^8 + x$. If we factor this polynomial into irreducible polynomials, we obtain

$$x^8 + x = x\,(x+1)\left(x^3 + x + 1\right)\left(x^3 + x^2 + 1\right).$$

The minimal polynomials of 0 and 1 are x and $x + 1$, respectively. We saw that $x^3 + x + 1$ is the minimal polynomial of α. The remaining elements $\alpha^2, \ldots, \alpha^6$ have minimal polynomial equal to one of these two cubics. To see, for example, which is the minimal polynomial of α^2, we simply evaluate one at α^2. Trying the first cubic, we have;

$$\begin{aligned}(\alpha^2)^3 + \alpha^2 + 1 = \alpha^6 + \alpha^2 + 1 &= (\alpha^2 + 1) + \alpha^2 + 1 \\ &= 0;\end{aligned}$$

we used that $x^6 + I = x^2 + 1 + I$ to see that $\alpha^6 = \alpha^2 + 1$. Thus, $m_{\alpha^2}(x) = x^3 + x + 1$. Similarly, if we try α^3, we have

$$\begin{aligned}(\alpha^3)^3 + \alpha^3 + 1 = \alpha^9 + \alpha^3 + 1 &= \alpha^2 + \alpha^3 + 1 \\ &= \alpha^2 + (\alpha + 1) + 1 = \alpha^2 + \alpha \neq 0\end{aligned}$$

since $\alpha^7 = 1$ and $\alpha^3 + \alpha + 1 = 0$. Therefore, $m_{\alpha^3}(x) = x^3 + x^2 + 1$, the only possible choice since α^3 is not a root of the other three factors of $x^8 + x$. If we were to finish this calculation, we would see that $x^3 + x + 1 = m_\alpha(x) = m_{\alpha^2}(x) = m_{\alpha^4}(x)$ and $x^3 + x^2 + 1 = m_{\alpha^3}(x) = m_{\alpha^5}(x) = m_{\alpha^6}(x)$. By making a simple but quite helpful observation, we can make the process of finding minimal polynomials easier. To help motivate this result, we note that $x^3 + x + 1$ is the minimal polynomial of α, α^2, and α^4, and $x^3 + x^2 + 1$ is the minimal polynomial of α^3, $\alpha^6 = (\alpha^3)^2$, and $\alpha^5 = (\alpha^3)^4$.

To help us prove the following proposition, we note that if $a, b \in \mathrm{GF}(2^n)$, then $(a + b)^2 = a^2 + 2ab + b^2 = a^2 + b^2$ since $2c = c + c = 0$ for any $c \in \mathrm{GF}(2^n)$. Thus, we have the formula $(a + b)^2 = a^2 + b^2$. By an induction argument, we see that $(a_1 + \cdots + a_r)^2 = a_1^2 + \cdots + a_r^2$ for any r and any $a_i \in \mathrm{GF}(2^n)$.

Proposition 7.21. *Let $f(x) \in \mathbb{Z}_2[x]$. If $\alpha \in \mathrm{GF}(2^n)$ is a root of $f(x)$, then α^2 is also a root of $f(x)$.*

Write $f(x) = b_0 + \cdots + b_{r-1}x^{r-1} + x^r$. Note that each $b_i \in \mathbb{Z}_2 = \{0, 1\}$. Since $f(\alpha) = 0$, we have $b_0 + b_1\alpha + \cdots + b_{r-1}\alpha^{r-1} + \alpha^r$. Therefore,

$$f(\alpha^2) = \sum_i b_i(\alpha^2)^i = \sum_i b_i^2\alpha^{2i} = \left(\sum_i b_i\alpha^i\right)^2 = f(\alpha)^2 = 0.$$

Going back to the previous example, since $m_\alpha(x) = x^3 + x + 1$ has α as a root, the proposition implies that α^2 and $\alpha^4 = (\alpha^2)^2$ are roots as well. Their minimal polynomials must also be $x^3 + x + 1$ since the minimal polynomial of each divides $x^3 + x + 1$, and $x^3 + x + 1$ is irreducible over $\mathbb{Z}_2$, i.e., it has no divisors other than itself and 1. Similarly, once we know that $m_{\alpha^3}(x) = x^3 + x^2 + 1$, we conclude that this is also the minimal polynomial of $(\alpha^3)^2 = \alpha^6$ and $(\alpha^3)^4 = \alpha^5$. However, $(\alpha^3)^8 = (\alpha^5)^2 = \alpha^{10} = \alpha^3$, so we do not produce any more roots of $x^3 + x^2 + 1$. By the way, this shows that over GF(8), we have the factorizations

$$x^3 + x + 1 = (x - \alpha)(x - \alpha^2)(x - \alpha^4),$$
$$x^3 + x^2 + 1 = (x - \alpha^3)(x - \alpha^6)(x - \alpha^5).$$

Thus,

$$\begin{aligned} x^8 + x &= x(x + 1)(x^3 + x + 1)(x^3 + x^2 + 1) \\ &= (x - 0)(x - 1)(x - \alpha)(x - \alpha^2)(x - \alpha^3)(x - \alpha^4)(x - \alpha^5)(x - \alpha^6). \end{aligned}$$

In other words, $x^8 = x$ has one linear factor for each element of GF(8) as must be the case from the Remainder Theorem and its consequences.

Exercises

The following fact will be useful in some of the exercises: It is proved in a more advanced algebra course that given any field F there is a smallest field extension K of F in which every polynomial $p(x)$ in $F[x]$ factors into linear factors, i.e., every root of $p(x)$ lies in K. Such a field extension is called an algebraic closure of F. For example, $\mathbb{C}$ is an algebraic closure of $\mathbb{R}$.

1. Prove Proposition 7.5.
2. Let K be an algebraic closure of $\mathbb{Z}_2$. For $m > 0$ consider the polynomial $p_m(x) = x^{2^m} + x$ with coefficients in $\mathbb{Z}_2$ that arose in Theorem 7.18. Prove that the collection of roots in K of $p_m(x)$ is itself a field extension of $\mathbb{Z}_2$. For example, if α and β are roots of $p_m(x)$, then $(\alpha + \beta)^{2^m} = \alpha^{2^m} + \beta^{2^m} = \alpha + \beta$ so the collection of roots of $p_m(x)$ is closed under addition.
   In fact $p_m(x)$ has exactly 2^m distinct roots. In this way we obtain field extensions of $\mathbb{Z}_2$ of every possible dimension. The next few exercises provide a demonstration of this fact. In Exercises 3–6 let F be any field.
3. Let $f \in F[x]$. Writing $f(x) = \sum_{i=0}^{n} \alpha_i x^i$, define the formal derivative of f to be $f'(x) = \sum_{i=1}^{n} i\alpha_i x^{i-1}$. Prove that formal differentiation satisfies the sum and product rules: $(f + g)' = f' + g'$ and $(fg)' = f'g + fg'$. The chain rule also holds. Note that if $f = x^2 \in \mathbb{Z}_2[x]$ then $f' = 0$, so unlike in calculus, polynomials of 0 derivative don't necessarily represent constant functions.
4. Suppose that K is an algebraic closure of F. Let $f \in F[x]$ and suppose that $\alpha \in K$ is a root of multiplicity greater than one of f, meaning that $(x - \alpha)^2$ divides f in $K[x]$. Prove that α is also a root of f'. In particular, show that as polynomials in $K[x]$ the $\gcd(f, g) \neq 1$.
5. Let L be any extension field of F. A pair of polynomials f and g in $F[x]$ can also be viewed as lying in $L[x]$. Use Proposition 5.5 to see that if d and e are the gcds of f and g computed over

F and L respectively, then each one divides the other. Thus $\gcd(f, g)$ is well defined in the sense that it is the same polynomial in $F[x]$ whether f and g are viewed in $F[x]$ or $L[x]$.
6. Prove that the polynomial $f \in F[x]$ has no multiple roots if and only if f and f' are relatively prime.
7. Prove that the collection of roots of the polynomial $p_m(x)$ of Exercise 2 above is a field with exactly 2^m elements.
8. Let $F = \text{GF}(8)$ as in Example 7.22. There are three roots of the polynomial $x^3 + x^2 + 1$ in F; find them. Choose one of the roots; call it t. Calculate t, t^2, and t^4 to see that t, t^2, t^4 are the three roots of $x^3 + x^2 + 1$.
9. Let $F = \text{GF}(8)$ again. There are three roots of the polynomial $x^3 + x + 1$ in F; find them. Choose one of the roots; call it t. Calculate t, t^2, and t^4 to see that t, t^2, t^4 are the three roots of $x^3 + x + 1$.
10. Let α be a root of the polynomial $x^4 + x + 1 \in \mathbb{Z}_2[x]$. For each $0 \leq i \leq 15$ express α^i as a $\mathbb{Z}_2$ linear combination of $1, \alpha, \alpha^2, \alpha^3$ verifying that α is a primitive element for GF(16).

7.4 Reed–Solomon Codes

The music that you hear on a CD and the information in QR codes (the increasingly common two-dimensional bar codes readable by smart phones) are encoded using a special class of cyclic codes known as Reed–Solomon codes. As we saw in Sect. 7.1, we can produce a cyclic code of length m by finding a polynomial $g(x)$ which divides $x^m + 1$. We use this idea to produce important codes of length $2^n - 1$ using a primitive element for $GF(2^n)$. Choose a positive integer n, and let t be an integer with $t \leq 2^{n-1} - 1$. For α a primitive element of $\text{GF}(2^n)$, let $g(x)$ be the least common multiple of the minimal polynomials of the elements $\alpha, \alpha^2, \ldots, \alpha^{2t}$ (i.e., the polynomial of least degree with these roots). The code whose generator polynomial is $g(x)$ is called a Reed–Solomon code with *designated distance* $d = 2t + 1$. We determine the dimension of this code and see why d is called the designated distance in the theorem below. We denote the code constructed in this way $RS(2^n - 1, t, \alpha)$, since its length is $2^n - 1$ and the code depends on t and the choice of a primitive element α of $\text{GF}(2^n)$.

Example 7.22. Let $n = 3$, let α be a primitive element of GF(8), and let $C = RS(7, 1, \alpha)$. This means that the generator polynomial $g(x)$ is the minimal polynomial of α since this polynomial will vanish on α and α^2 Thus $\deg(g(x)) = 3$, so the dimension k of C is $2^n - 1 - \deg(g(x)) = 7 - 3 = 4$. To be more explicit, let us represent GF(8) as $\mathbb{Z}_2[x]/(x^3 + x + 1)$ and choose α to be the coset of x. Then $g(x) = x^3 + x + 1$. From our association between polynomials and words, the basis $[g(x), xg(x), x^2g(x), x^3g(x)]$ corresponds to the four words $[1101000, 0110100, 0011010, 0001101]$. A generator matrix for C is then

$$\begin{pmatrix} 1\,1\,0\,1\,0\,0\,0 \\ 0\,1\,1\,0\,1\,0\,0 \\ 0\,0\,1\,1\,0\,1\,0 \\ 0\,0\,0\,1\,1\,0\,1 \end{pmatrix}.$$

If we compute a parity check matrix for C, we can get

$$H = \begin{pmatrix} 1\,0\,1\,1\,1\,0\,0 \\ 1\,1\,1\,0\,0\,1\,0 \\ 0\,1\,1\,1\,0\,0\,1 \end{pmatrix}$$

to be such a matrix. Looking at H, we see that it closely resembles the Hamming matrix; in fact, it is obtained from the Hamming matrix by appropriately rearranging the columns. This Reed–Solomon code is obtained from the (7,4) Hamming code by a permutation of the entries of the codewords. Recall that the (7,4) Hamming code corrects one error.

Theorem 7.23. *Let $C = RS(2^n - 1, t, \alpha)$. Then the distance d of C is at least $2t + 1$ so that C corrects at least t errors.*

Proof. Set $m = 2^n - 1$. The code C consists of all cosets in $Z_2[x]/(x^m + 1)$ of polynomials with roots $\alpha, \alpha^2, \ldots, \alpha^{2t}$. According to our convention that identifies polynomials of degree strictly less than m with the vector of their coefficients (of length m), C is the nullspace of the $2t \times m$ matrix

$$H = \begin{pmatrix} 1 & \alpha & \alpha^2 & \ldots & \alpha^{m-1} \\ 1 & \alpha^2 & (\alpha^2)^2 & \ldots & (\alpha^2)^{m-1} \\ 1 & \alpha^3 & (\alpha^3)^2 & \ldots & (\alpha^3)^{m-1} \\ \cdot & \cdot & \cdot & \cdots & \cdot \\ \cdot & \cdot & \cdot & \cdots & \cdot \\ \cdot & \cdot & \cdot & \cdots & \cdot \\ 1 & \alpha^{2t} & (\alpha^{2t})^2 & \ldots & (\alpha^{2t})^{m-1} \end{pmatrix}.$$

The distance of C is at least d if every set of $d - 1 = 2t$ columns of H is linearly independent (i.e., if their sum is nonzero). This condition holds if and only if every submatrix of H consisting of $2t$ columns has nonzero determinant. To see that this is satisfied, consider such a submatrix:

$$M = \begin{pmatrix} \alpha^{j_1} & \alpha^{j_2} & \ldots & \alpha^{j_{2t}} \\ (\alpha^{j_1})^2 & (\alpha^{j_2})^2 & & (\alpha^{j_{2t}})^2 \\ (\alpha^{j_1})^3 & (\alpha^{j_2})^3 & & \\ \cdot & \cdot & \cdots & \cdot \\ \cdot & \cdot & \cdots & \cdot \\ & \cdot & \cdots & \cdot \\ (\alpha^{j_1})^{2t} & (\alpha^{j_2})^{2t} & & (\alpha^{j_{2t}})^{2t} \end{pmatrix}.$$

The following lemma shows, after factoring out α^{j_i} from the ith column, that $\det(M) = (\alpha^{j_1}\alpha^{j_2}\cdots\alpha^{j_{2t}})\Pi$ where Π is the product of all $(\alpha^{j_i} - \alpha^{j_k})$ where $j_i > j_k$. In particular, since $2t \leq m < 2^n$, $\alpha^{j_i} - \alpha^{j_k} \neq 0$ for each such j_i, j_k, we see that $\Pi \neq 0$. □

Lemma 7.24 (Vandermonde Determinant). *Let F be a field and $a_1, a_2, \ldots, a_n$ distinct elements of F. The determinant of the Vandermonde matrix*

$$V = \begin{pmatrix} 1 & 1 & \ldots & 1 \\ a_1 & a_2 & \ldots & a_n \\ \cdot & \cdot & \cdots & \cdot \\ \cdot & \cdot & \cdots & \cdot \\ \cdot & \cdot & \cdots & \cdot \\ a_1^{n-1} & a_2^{n-1} & \ldots & a_n^{n-1} \end{pmatrix}$$

is equal to $(a_n - a_1)(a_n - a_2)\cdots(a_n - a_{n-1})\cdot(a_{n-1} - a_1)\cdots(a_{n-1} - a_{n-2})\cdots(a_1 - a_1) = \Pi_{i>j}(a_i - a_j)$.

Proof. Argue by induction on n beginning with $n = 2$:

$$\det\begin{pmatrix} 1 & 1 \\ a_1 & a_2 \end{pmatrix} = a_2 - a_1.$$

Suppose that the result holds for n elements of F, $n \geq 2$. Consider the matrix

$$M(x) = \begin{pmatrix} 1 & 1 & \dots & 1 & 1 \\ a_1 & a_2 & \dots & a_n & x \\ . & . & \dots & . & \\ . & . & \dots & . & \\ . & . & \dots & . & \\ a_1^{n-1} & a_2^{n-1} & \dots & a_n^{n-1} & x^{n-1} \\ a_1^n & a_2^n & \dots & a_n^n & x^n \end{pmatrix}$$

Note that $M(a_{n+1})$ is the Vandermonde matrix corresponding to $a_1, a_2, \dots, a_{n+1}$. The determinant of $M(x)$ is a polynomial of degree precisely n with the n distinct roots $a_1, a_2, \dots, a_n$. By repeated use of the Remainder Theorem, $\det(M(x)) = \alpha \Pi_{i=1}^n (x - a_i)$ where α is the coefficient of x^n. But the coefficient of x^n is exactly the determinant of the Vandermonde matrix

$$\begin{pmatrix} 1 & 1 & \dots & 1 \\ a_1 & a_2 & \dots & a_n \\ . & . & \dots & . \\ . & . & \dots & . \\ . & . & \dots & . \\ a_1^{n-1} & a_2^{n-1} & \dots & a_n^{n-1} \end{pmatrix}$$

which, by the induction hypothesis, is equal to $\Pi_{i>j}(a_i - a_j)$. Thus $\det(M(x)) = \Pi_{i=1}^n (x - a_i)\Pi_{i>j}(a_i - a_j)$ and evaluation at $x = a_{n+1}$ completes the proof. □

Example 7.25. Let's build a 3 error correcting code of length 15, i.e., $RS(15, 3, \alpha)$ where $\alpha = \overline{x} \in \mathbb{Z}_2[x]/(x^4 + x + 1)$ is a primitive element for this concrete realization of $GF(16)$. The objective is to find the polynomial $g \in \mathbb{Z}_2[x]$ of least degree with roots $\alpha, \alpha^2, \dots, \alpha^6$. Note that $m_\alpha(x) = x^4 + x + 1$ so that $x^4 + x + 1$ divides g. Recalling that if $\beta \in GF(16)$ and $f \in \mathbb{Z}_2[x]$ has β as a root then β^{2^i} is a root of f for every i, as a root we see that α^2 and α^4 are also roots of $m_\alpha(x)$.

Now we must determine $m_{\alpha^3}(x), m_{\alpha^5}(x)$, and $m_{\alpha^6}(x)$. Arguments similar to those above show that $h(x) = (x - \alpha^3)(x - \alpha^6)(x - \alpha^9)(x - \alpha^{12})$ divides $m_{\alpha^3}(x)$ and therefore g (note that $\alpha^{15} = 1$ so that $\alpha^{24} = \alpha^9$). Thus this product must equal $m_{\alpha^3}(x) = m_{\alpha^6}(x)$ if it lies in $\mathbb{Z}_2[x]$. One can either multiply out the factors of h and check coefficients, or argue as follows. Since the degree of h is four, $h(x) = m_{\alpha^3}(x)$ if there is an irreducible polynomial of degree 4 in $\mathbb{Z}_2[x]$ that has α^3 as a root. Of the polynomials of degree 4 in $\mathbb{Z}_2[x]$, only $x^4 + x + 1, x^4 + x^3 + 1$, and $x^4 + x^3 + x^2 + x + 1$ are irreducible (check this!). Evaluating the last polynomial at α^3 reveals that $x^4 + x^3 + x^2 + x + 1 = m_{\alpha^3}(x)$.

Finally we determine $m_{\alpha^5}(x)$. Consider $(x - \alpha^5)(x - \alpha^{10})$ noting that $\alpha^{20} = \alpha^5$. We see that $(x - \alpha^5)(x - \alpha^{10}) = x^2 + (\alpha^5 + \alpha^{10})x + 1 = x^2 + x + 1 = m_{\alpha^5}(x)$. Thus $g(x) = (x^4+x+1)(x^4+x^3+x^2+x+1)(x^2+x+1)$ generates a 3 error correcting code of dimension 5.

Remark 7.26. Reed–Solomon codes are a special subclass of the cyclic codes known as BCH codes. Just as with Reed–Solomon codes, BCH codes are built with a primitive element α for

$GF(2^n)$, but instead of having length 2^n-1, a general BCH code can have as length any divisor m of 2^n-1. Choosing $\beta = \alpha^{\frac{2^n-1}{m}} \in GF(2^n)$, and some $i > 0$, the generator polynomial for a BCH code correcting t errors vanishes on $\beta^i, \beta^{i+1}, \ldots, \beta^{i+2t-1}$.

Exercises

Exercises 1–3 refer to the $RS(15, 3, \alpha)$ code constructed above.

1. Write down any four codeword polynomials in this code.
2. Determine whether or not $x^{13} + x^{12} + x^{11} + x^{10} + x^6 + x^3 + x^2 + 1$ is a codeword polynomial.
3. Check that $r(x) = x^{13} + x^{12} + x^{11} + x^{10} + x^9 + x^8 + x^5 + x^4 + x^3 + x + 1$ is not a codeword. Calculate the $S_i = r(\alpha^i)$ and write them as powers of α. Next, obtain the $2t \times 2t$ matrix A whose i, j entry is S_{i+j-1}. Determine the number of errors v by finding the largest l for which the $l \times l$ top left part of A is invertible. Use Section-7.4-Exercise-3.mw to help you with these computations.
4. Consider the $RS(15, 2, \alpha)$ code determined by $\alpha = \overline{x} \in \mathbb{Z}_2[x]/(x^4 + x + 1)$. Find the generator polynomial $g(x)$ (its degree is 8) and write down two nonzero codewords. Change each by making two errors. Run through, for each word, the steps you took for the previous problem, and see that you recover the original codeword.
5. Using the code of the previous problem, take a valid codeword and change it by making three errors. Go through the decoding procedure for it. Do you recover the original codeword?
6. A fact about RS codes is that if g is the generator polynomial for an RS code defined with a primitive element for $GF(2^n)$ and $m = \deg(g)$, then the code has dimension $2^n - 1 - m$. Determine the number of codewords in the codes of Exercises 1 and 4.
7. If you wish to transmit color pictures using 2,048 colors, you will need 2,048 codewords. If you wish to do this with an RS code using the field F of Exercise 1, what degree of a generator polynomial will you need to use? By using the generator procedure, find the largest value of t for which the corresponding code has at least 2,048 codewords (recall that $t \leq n/2$ in order for the code to correct at least t errors, so do not consider values of t larger than $n/2$). What percentage of errors in a codeword can be corrected?
8. Continuing the previous problem, if you instead use $F = \mathbb{Z}_2[x]/(x^5 + x^2 + 1)$, determine the length of an RS code you obtain from this field, and determine the degree of a generator polynomial needed to have 2,048 codewords. What is the largest value of t you can use to have 2,048 codewords? By using this larger field, can you correct a larger percentage of errors in a codeword than in the previous problem?

7.5 Error Correction for Reed–Solomon Codes

In this section a procedure is developed for the correction of any set of t or fewer errors for an $RS(2^n - 1, t, \alpha)$ code. Using our convention identifying words with cosets of polynomials of degree strictly less than $2^n - 1$, we'll consider only polynomial expressions. Thus, we suppose that $c(x)$ is a codeword that is received as $r(x)$. If $r(x) \neq c(x)$, write $r(x) = c(x) + e(x)$, where $e(x)$ is the error polynomial. Note that $e(x) = x^{m_1} + x^{m_2} + \cdots + x^{m_k}$ where k is equal to the number of errors that have occurred in positions $m_1, m_2, \ldots, m_k$. Our decoding procedure will first determine k and then locate the m_j, thus determining $e(x)$. The codeword $c(x)$ is then recovered as $r(x) + e(x)$.

By the construction of the RS code, $c(\alpha^i) = 0$ for $i = 1, 2, \ldots, 2t$. This already gives us some information about $e(x)$, namely, for $i = 1, 2, \ldots, 2t$ we have

$$\begin{aligned} r(\alpha^i) &= c(\alpha^i) + e(\alpha^i) \\ &= e(\alpha^i). \end{aligned}$$

In particular, the $2t$ elements of $GF(2^n)$ given by

$$e(\alpha^i) = (\alpha^{m_1})^i + (\alpha^{m_2})^i + \ldots + (\alpha^{m_k})^i$$

are known, and from these we will recover k and the m_j. To that end, for each $j = 1, \ldots, k$ set Y_j equal to the as yet unknown powers α^{m_j} of our primitive element for $GF(2^n)$, so that

$$\begin{aligned} e(\alpha) &= \sum_{j=1}^{k} Y_j, \\ e(\alpha^2) &= \sum_{j=1}^{k} Y_j^2, \\ &\vdots \\ e(\alpha^{2t}) &= \sum_{j=1}^{k} Y_j^{2t}. \end{aligned}$$

To simplify the notation and to emphasize the relationship with the Y_j, set

$$e(\alpha^i) = S_i = S_i(Y_1, \ldots, Y_k) = \sum_{j=1}^{k} Y_j^i.$$

Thus, our goal is to determine k and the Y_j by constructing a polynomial in $GF(2^n)$ whose roots are precisely the Y_j. Solving for the Y_j as powers of α will give the positions of the errors.

Certainly $L(z) = (z + Y_1)(z + Y_2) \cdots (z + Y_k)$ is the desired polynomial, called the error locator polynomial, but we don't as yet know its coefficients. We can, however, calculate them from the S_i together with a version of a result known as Newton's identities. Collecting coefficients of powers of z we have

$$L(z) = \sum_{i=1}^{k} \sigma_{k-i}(Y_1, \ldots, Y_k) z^i$$

where $\sigma_0 = 1$,

$$\sigma_1(Y_1, \ldots, Y_k) = Y_1 + \cdots + Y_k = \sum_{j=1}^{k} Y_j,$$

$$\sigma_2(Y_1,\ldots,Y_k) = Y_1^2 + Y_1Y_2 + \cdots + Y_1Y_k + \cdots + Y_{k-1}Y_k + Y_k^2 = \sum_{1\le i\le j\le k} Y_iY_j,$$

$$\sigma_3(Y_1,\ldots,Y_k) = \sum_{1\le i\le j\le r\le k} Y_iY_jY_r, \text{ and}$$

$$\vdots$$

$$\sigma_k(Y_1,\ldots,Y_k) = Y_1Y_2\cdots Y_k$$

are known as the elementary symmetric functions in $Y_1, Y_2, \ldots, Y_k$. The (to be determined) σ_i are related to the (known) S_i by Newton's identities:

Lemma 7.27. *For each $i = 1, 2, \ldots, 2t$ we have*

$$S_i\sigma_k + S_{i+1}\sigma_{k-1} + \cdots + S_{i+k-1}\sigma_1 + S_{i+k} = 0.$$

Proof. For each $j = 1, \ldots, k$, evaluate $L(z) = \sum_{i=1}^{k} \sigma_{k-i}(Y_1,\ldots,Y_k)z^i$ at $z = Y_j$ to obtain

$$0 = Y_j^k + \sigma_1 Y_j^{k-1} + \sigma_2 Y_j^{k-2} + \cdots + \sigma_{k-1}Y_j + \sigma_k.$$

Multiplying each equation by Y_j^i results in

$$0 = Y_j^{i+k} + \sigma_1 Y_j^{i+k-1} + \sigma_2 Y_j^{i+k-2} + \cdots + \sigma_{k-1}Y_j^{i+1} + \sigma_k Y_j^i$$

and summing these j equations yields

$$0 = \sum_{j=1}^{k} Y_j^{i+k} + \sigma_1 \sum_{j=1}^{k} Y_j^{i+k-1} + \sigma_2 \sum_{j=1}^{k} Y_j^{i+k-2} + \cdots + \sigma_{k-1}\sum_{j=1}^{k} Y_j^{i+1} + \sigma_k \sum_{j=1}^{k} Y_j^i,$$

which is the desired result. □

Recalling once again that we are working over $\mathbb{Z}_2$, view the system of equations $S_i\sigma_k + S_{i+1}\sigma_{k-1} + \cdots + S_{i+k-1}\sigma_1 = S_{i+k}$ for $1 \le i \le k$ as a linear system in the k unknowns σ_j with coefficients S_{i+k-j}. In matrix form this becomes

$$\begin{pmatrix} S_1 & S_2 & \ldots & S_k \\ S_2 & S_3 & & S_{k+1} \\ \cdot & \cdot & \ldots & \cdot \\ S_k & S_{k+1} & \ldots & S_{2k-1} \end{pmatrix} \begin{pmatrix} \sigma_k \\ \sigma_{k-1} \\ \cdot \\ \cdot \\ \cdot \\ \sigma_1 \end{pmatrix} = \begin{pmatrix} S_{k+1} \\ S_{k+2} \\ \cdot \\ S_{2k} \end{pmatrix}. \qquad (\bigstar)$$

If the matrix of coefficients is invertible we solve it for the σ_j, which are the coefficients in the error locator polynomial $L(z)$, then solve $L(z)$. Expressing the roots of $L(z)$ as powers of α locates the positions of the errors.

All of this development is done under the assumption that k errors have been made. But as of yet we don't know the value of k. The next lemma enables us to determine k.

Lemma 7.28. *Let $\mathbb{S}$ be the $t \times t$ matrix*

$$\begin{pmatrix} S_1 & S_2 & \dots & S_t \\ S_2 & S_3 & & S_{t+1} \\ \cdot & \cdot & \dots & \cdot \\ S_t & S_{t+1} & \dots & S_{2t-1} \end{pmatrix}.$$

Then k is the largest integer for which the $k \times k$ submatrix

$$\mathbb{S}_k = \begin{pmatrix} S_1 & S_2 & \dots & S_k \\ S_2 & S_3 & & S_{k+1} \\ \cdot & \cdot & \dots & \cdot \\ S_k & S_{k+1} & \dots & S_k \end{pmatrix}$$

is invertible.

Thus to find k begin with the matrix $\mathbb{S}$. If $\mathbb{S}$ is invertible, then $k = t$. Otherwise, delete the last column and last row of $\mathbb{S}$ and check for invertibility. Iteration of this process will result in an invertible matrix $\mathbb{S}_k$ and hence the linear system ★ can be solved for the σ_j.

Proof (Of the lemma). We show that the matrix $\mathbb{S}_k$ is nonsingular while if $l > k$ the matrix $\mathbb{S}_l$ is singular. Note that $l \leq t$. Recalling that $S_i = \sum_{j=1}^{k} Y_j^i$, the matrix $\mathbb{S}_l$ factors as

$$\underbrace{\begin{pmatrix} S_1 & S_2 & \dots & S_l \\ S_2 & S_3 & & S_{l+1} \\ \cdot & \cdot & \dots & \cdot \\ S_l & S_{l+1} & \dots & S_l \end{pmatrix}}_{\mathbb{S}_l} =$$

$$\underbrace{\begin{pmatrix} 1 & 1 & \dots & 1 \\ Y_1 & Y_2 & \dots & Y_l \\ \cdot & \cdot & \dots & \cdot \\ Y_1^{l-1} & Y_2^{l-1} & \dots & Y_l^{l-1} \end{pmatrix}}_{A} \underbrace{\begin{pmatrix} Y_1 & 0 & \dots & 0 \\ 0 & Y_2 & 0 \dots & 0 \\ \cdot & \cdot & \dots & \cdot \\ 0 & 0 & \dots & Y_l \end{pmatrix}}_{D} \underbrace{\begin{pmatrix} 1 & Y_1 & Y_1^2 & \dots & Y_1^{l-1} \\ 1 & Y_2 & Y_2^2 & \dots & Y_2^{l-1} \\ \cdot & \cdot & \cdot & \dots & \cdot \\ 1 & Y_2 & \cdot & \dots & Y_2^{l-1} \end{pmatrix}}_{A^T}.$$

Moreover, the entries of $\mathbb{S}_l$ depend only on $Y_1, \dots, Y_k$. Thus, if $l > k$, we may take $Y_{k+1} = Y_{k+2} = \dots = Y_l = 0$ to obtain that D, and therefore $\mathbb{S}_l$, is singular. The Y_j for $1 \leq j \leq k$ are all nonzero though, and furthermore distinct, as they are distinct powers of the primitive element α. Moreover, A is a Vandermonde matrix, so that with $|M|$ denoting the determinant of a square matrix M, we have

$$\begin{aligned} |\mathbb{S}_k| &= |A|\,|D|\,|A^T| = |A|^2\,|D| \\ &= \Pi_{1 \leq i \leq j \leq k}(Y_i - Y_j)^2 \Pi_{i=1}^{k} Y_i \neq 0. \end{aligned}$$

□

Example 7.29. Let's see how the decoding procedure is implemented on the $RS(15, 3, \alpha)$ constructed above. Suppose that the received word is the polynomial $r(x) = x + x^2 + x^3 +$

$x^4 + x^5 + x^6 + x^7 + x^{10} + x^{12}$. We compute the $S_i = r(\alpha^i)$, $i = 1, 2, 3, 4, 5 = 2t - 1$ using the relations $\alpha^4 = \alpha + 1$, $\alpha^{15} = 1$.

$$S_1 = \alpha + \alpha^2 + \alpha^3 + \alpha^4 + \alpha^5 + \alpha^6 + \alpha^7 + \alpha^{10} + \alpha^{12} = \alpha^2$$

$$S_2 = \alpha^2 + \alpha^4 + \alpha^6 + \alpha^8 + \alpha^{10} + \alpha^{12} + \alpha^{14} + \alpha^{20} + \alpha^{24} = \alpha + 1$$

$$S_3 = \alpha^3 + \alpha^6 + \alpha^9 + \alpha^{12} + \alpha^{15} + \alpha^{18} + \alpha^{21} + \alpha^{30} + \alpha^{36} = \alpha^3 + 1$$

$$S_4 = \alpha^4 + \alpha^8 + \alpha^{12} + \alpha^{16} + \alpha^{20} + \alpha^{24} + \alpha^{28} + \alpha^{40} + \alpha^{48} = \alpha^2 + 1$$

$$S_5 = \alpha^5 + \alpha^{10} + \alpha^{15} + \alpha^{20} + \alpha^{25} + \alpha^{30} + \alpha^{35} + \alpha^{50} + \alpha^{60} = 1$$

Consider the matrix

$$\mathbb{S} = \begin{pmatrix} S_1 & S_2 & S_3 \\ S_2 & S_3 & S_4 \\ S_3 & S_4 & S_5 \end{pmatrix} = \begin{pmatrix} \alpha^2 & \alpha + 1 & \alpha^3 + 1 \\ \alpha + 1 & \alpha^3 + 1 & \alpha^2 + 1 \\ \alpha^3 + 1 & \alpha^2 + 1 & 1 \end{pmatrix}.$$

Since $|\mathbb{S}| = 0$ there are fewer than 3 errors.

We next consider

$$\mathbb{S}_2 = \begin{pmatrix} S_1 & S_2 \\ S_2 & S_3 \end{pmatrix} = \begin{pmatrix} \alpha^2 & \alpha + 1 \\ \alpha + 1 & \alpha^3 + 1 \end{pmatrix}.$$

Since $|\mathbb{S}_2| = \alpha^2 + \alpha + 1 \neq 0$, there are exactly two errors, and we can solve the system

$$\mathbb{S}_2 \begin{pmatrix} \sigma_2 \\ \sigma 1 \end{pmatrix} = \begin{pmatrix} S_3 \\ S_4 \end{pmatrix}$$

for the coefficients of the error locator polynomial $L(z) = z^2 + \sigma_1 z + \sigma_2$. Our system

$$\begin{pmatrix} \alpha^2 & \alpha + 1 \\ \alpha + 1 & \alpha^3 + 1 \end{pmatrix} \begin{pmatrix} \sigma_2 \\ \sigma_1 \end{pmatrix} = \begin{pmatrix} \alpha^3 + 1 \\ \alpha^2 + 1 \end{pmatrix}$$

can be solved by inverting $\mathbb{S}_2$. Note that $\alpha^2 + \alpha + 1 = \alpha^{10}$, and $\alpha^{-10} = \alpha^5$ so

$$\mathbb{S}_2^{-1} = \alpha^5 \begin{pmatrix} \alpha^3 + 1 & \alpha + 1 \\ \alpha + 1 & \alpha^2 \end{pmatrix}.$$

The solutions $\sigma_1 = \alpha^2, \sigma_2 = \alpha^3 + \alpha^2$ are then obtained by matrix multiplication to obtain $L(z) = z^2 + \alpha^2 z + \alpha^3 + \alpha^2$. In a general situation, one could search through $GF(2^m)$ for the roots of $L(z)$, but in the present context observe that $L(z) = (z + \alpha)^2 + \alpha^2(z + \alpha)$ so that either $z = \alpha$ or $z = \alpha + \alpha^2 = \alpha^5$. Thus we have determined that the errors appear in positions 1 and 5 and we decode $r(x) = x + x^2 + x^3 + x^4 + x^5 + x^6 + x^7 + x^{10} + x^{12}$ to $x^2 + x^3 + x^4 + x^6 + x^7 + x^{10} + x^{12} = x^2 g$.

References

1. Hankerson D et al (2000) Coding theory and cryptography: the essentials, 2nd edn. Marcel Dekker, New York
2. van der Waerden BL (1971) Modern algebra. Springer, Berlin

Chapter 8
Groups and Cryptography

The final two applications of abstract algebra we will discuss are to cryptography, i.e., secure transmission of private information, and to the classification of geometric patterns in the plane $\mathbb{R}^2$. The algebraic structure at the heart of both applications is that of a group. The structure known as a group was perhaps the first algebraic structure to have been studied purely abstractly, and is one of the most fundamental of mathematical structures. As we shall see, the underlying additive structures of rings, fields, and vector spaces are all special examples of groups. Unlike these other structures, a group has only a single operation.

The earlier chapters demonstrate the value of "abstracting" from specific cases to a general framework. Hence we abstracted from simple identification numbers to modular arithmetic in order to understand error detection capability and develop more sensitive identification schemes. By abstracting the notion of redundancy in the check digit, we saw how to judiciously introduce more redundancies to actually correct errors. Further abstraction of the notion of distance led ultimately to algebraic extensions of finite fields, and codes with enhanced error detection capability. In a similar vein, abstractly algebraicizing number constructions in terms of field extensions of the rational numbers led to the solution of geometric questions dating to antiquity. This chapter will introduce the elements of group theory necessary for the most common contemporary method to encrypt electronic passwords.

8.1 Definition and Examples of Groups

To motivate the definition of a group, recall that a unit of a ring R is an element having a multiplicative inverse. We denote by R^* the set of all units of R. In other words,

$$R^* = \{a \in R : \text{there is a } c \in R \text{ with } ac = ca = 1\}.$$

Recall also that if a, b are units of R, then so is ab, since ab has $b^{-1}a^{-1}$ as its multiplicative inverse. Thus, if we multiply two elements of R^*, the result is another element of R^*. Multiplication then induces a binary operation on the set R^*. We note three properties of this binary operation:

Electronic supplementary material The online version of this chapter (doi:10.1007/978-3-319-04498-9_8) contains supplementary material, which is available to authorized users. The supplementary material can also be downloaded from http://extras.springer.com.

D.R. Finston and P.J. Morandi, *Abstract Algebra: Structure and Application*,
Springer Undergraduate Texts in Mathematics and Technology, DOI 10.1007/978-3-319-04498-9_8

multiplication on R^* is associative; $1 \in R^*$, so R^* has an identity; and each element of R^* has an inverse in R^*. It is these properties that make up the definition of a group.

Example 8.1. Let $R = \mathrm{M}_n(\mathbb{R})$, the ring of all $n \times n$ matrices with real number entries (R is a noncommutative ring if $n > 1$). Then R^* is the set of invertible (i.e., nonsingular) $n \times n$ matrices, equivalently $n \times n$ matrices of nonzero determinant. Recalling properties of matrix inversion and determinants, we have for nonsingular matrices A and B:

$$(AB)^{-1} = B^{-1}A^{-1}$$
$$\det(AB) = \det(A)\det(B),$$

demonstrating closure of R^* under multiplication.

Definition 8.2. Let G be a nonempty set together with a binary operation $*$ on G. Then the pair $(G, *)$ is said to be a group if

1. $a * (b * c) = (a * b) * c$ for all $a, b, c \in G$,
2. there is an $e \in G$ such that $e * a = a * e = a$ for all $a \in G$, and
3. for each $a \in G$ there is an element $b \in G$ with $a * b = b * a = e$.

Naturally enough, the first axiom is referred to as the Associative Property, the element e is referred to as the identity element of G, and the element b is referred to as the inverse of the element a. Uniqueness of the identity element and of inverses will be established below, justifying the use of the word "the."

While the group axioms seem abstract, there are many natural examples of groups; indeed, the motivating framework does yield a group:

Example 8.3. Let $(R, +.\cdot)$ be a ring. Then associativity of ring multiplication and the definition of unit shows that $(R^*, \cdot)$ is a group. The identity of R^* is the multiplicative identity 1 of R. For instance, with $R = \mathbb{Z}$, the group $\mathbb{Z}^*$ is nothing more than the set $\{1, -1\}$ under multiplication.

The example above with $R = \mathrm{M}_n(\mathbb{R})$ is of such importance that $\mathrm{M}_n(\mathbb{R})^*$ has its own name and notation.

Definition 8.4. The *General Linear Group* of $n \times n$ matrices with real number entries, denoted $\mathrm{GL}_n(\mathbb{R})$, is the group of units of $\mathrm{M}_n(\mathbb{R})$. Note that the operation, i.e., matrix multiplication, in $\mathrm{GL}_n(\mathbb{R})$ is noncommutative if $n > 1$.

More generally, if R is any commutative ring, the set $\mathrm{M}_n(R)$ of $n \times n$ matrices with entries in R is again a ring (noncommutative if $n > 1$). The group of units of $\mathrm{M}_n(R)$, denoted $\mathrm{GL}_n(R)$, is referred to as the General Linear Group of $n \times n$ matrices over R. The application to symmetry groups will make use of this group for $R = \mathbb{Z}$.

The first order of business is to prove the uniqueness of the identity and of inverses.

Proposition 8.5. *Let G be a group.*

1. *If $e * a = a * e = a$ for all $a \in G$ and $e' * a = a * e' = a$ for all $a \in G$, then $e = e'$.*
2. *If $a, b, b' \in G$ satisfy $a * b = b * a = a * b' = b' * a = e$, then $b = b'$.*

Proof. For assertion 1., taking $a = e'$, the hypothesis $e * a = a$ yields $e * e' = e'$. Taking $a = e$, the hypothesis $a * e' = a$ yields $e * e' = e$. Assertion 2. is a consequence of the associative property of the group operation.

$$b = b * e = b * (a * b') \\ = (b * a) * b' = e * b' = b'$$

□

Example 8.6. Let R be a ring. Then $(R, +)$ is a group. To see this, we remark that the definition of a ring includes the three properties of a group when considering addition. Note that the identity of $(R, +)$ is 0, the additive identity of R. Here the operation is of course commutative.

The two examples $(R, +)$ and $(R^*, \cdot)$ illustrate a potential source of notational difficulty. In the first example, the group operation is multiplication, so it is natural to write, for n a positive integer, $\underbrace{g \cdot g \cdot \cdots \cdot g}_{n \text{ times}}$ as g^n and $\underbrace{g^{-1} \cdot g^{-1} \cdot \cdots \cdot g^{-1}}_{n \text{ times}}$ as g^{-n}. In the second example, the natural notations are ng for $\underbrace{g + g + \cdots + g}_{n \text{ times}}$ and $-ng$ for $\underbrace{-g - g - \cdots - g}_{n \text{ times}}$. When writing about general groups, it is most common to write the operation multiplicatively. However, the context in which the group operations arise should make clear whether multiplicative or additive notation is most appropriate.

To formally define g^n, we first define $g^0 = e$, the identity of G. If n is a positive integer, then we define, inductively, $g^{n+1} = g^n \cdot g$. Therefore, $g^1 = g^0 \cdot g = e \cdot g = g$, and $g^2 = g \cdot g$, and so on. For negative exponents, if $n > 0$, we set $g^{-n} = (g^n)^{-1}$. The laws of exponents are consequences of the definition of a group (these are left as homework problems): $g^n \cdot g^m = g^{n+m}$, $(g^n)^m = g^{nm}$, and $(g^n)^{-1} = (g^{-1})^n$ for any $n, m \in \mathbb{Z}$. (Their additive analogs are: $ng + mg = (n+m)g$, $nmg = n(mg)$, and $-(ng) = -ng$).

It is standard in discussions about groups in general to use multiplicative notation for the operation, unless we are discussing a particular example or class of examples in which the operation is presented as addition (for instance, as $(R, +)$ for R a ring). In particular, we will write gh or $g \cdot h$ for the product of g and h, and g^{-1} for the inverse of g. Unless the symbol for the operation needs to be specified, we will refer to a group by a single symbol, such as G, rather than by a pair such as $(G, *)$. Despite this, it is important to remember that a group is not just a set but is a set together with a binary operation.

The special cases of $(\mathbb{Z}, +)$ and $(\mathbb{Z}_n, +)$ illustrate an important class of groups.

Definition 8.7. A group G is said to be cyclic if there is an element $g \in G$ so that each element of G has the form g^n for some $n \in \mathbb{Z}$ when $*$ written multiplicatively (or ng if $*$ is written additively). In this case, the element g is called a generator of G.

Example 8.8. The multiplicative group $\mathbb{Z}^* = \{1, -1\}$, since these are the only integers with multiplicative inverses. This group is cyclic with generator -1, while the groups $(\mathbb{Z}, +)$ and $(\mathbb{Z}_n, +)$ are cyclic (additive) groups with generators 1 and $\overline{1}$, respectively. Every group with two elements, say $G = \{a, e\}$, with e the identity, is cyclic generated by a.

Example 8.9. Every group with three elements is cyclic. Let $G = \{a, b, e\}$ with e the identity element. The only possibility for the product ab is e since multiplication by a^{-1} of both sides of the equation $ab = a$ would yield the contradiction $b = e$, and multiplication of $ab = a$ by b^{-1} would force $a = e$. It follows that $a^2 = b$ since $a^2 \neq e$ by uniqueness of the inverse, and again multiplication by a^{-1} on $a^2 = a$ would force $a = e$. Thus G is generated by a (and also by b by a symmetric argument).

Remark 8.10. In fact every group with a prime number of elements is cyclic as follows from Lagrange's Theorem below (Exercise 11).

Example 8.11. Let V be a vector space. Ignoring scalar multiplication and only considering addition, we see that $(V, +)$ is a group.

Example 8.12. For more special cases of groups extracted from rings, the following are groups under addition: $(\mathbb{Q}, +)$ and $(\mathbb{R}, +)$, while $(\mathbb{Q}^*, \cdot)$, and $(\mathbb{R}^*, \cdot)$ are groups under multiplication. In contrast with the finite group $\mathbb{Z}^* = \{1, -1\}$, we have the infinite groups $\mathbb{Q}^* = \mathbb{Q} - \{0\}$ and $\mathbb{R}^* = \mathbb{R} - \{0\}$ (they are groups because $\mathbb{Q}$ and $\mathbb{R}$ are fields so that every nonzero element has a multiplicative inverse in $\mathbb{Q}$ or $\mathbb{R}$, respectively).

Generalizing this observation we have:

Example 8.13. Let F be a field. Then $F^* = F - \{0\}$ is a group under multiplication. This is because F is a ring and $F^* = F - \{0\}$ by the definition of a field.

Example 8.14. Let $F = \text{GF}(2^n)$. The existence of a primitive element for F says exactly that F^* is a cyclic group.

Example 8.15. Let X be any set. We denote by $P(X)$ the set of all *permutations* of X. That is,

$$P(X) = \{f : X \to X : f \text{ is a 1-1, onto function}\} .$$

We define an operation on $P(X)$ by composition of functions; note that if f, g are functions from X to X, then so is $f \circ g$, and that if f and g are both 1-1 and onto, then so is $f \circ g$. Therefore, composition is indeed a binary operation on $P(X)$. We leave it as an exercise to show that $P(X)$ is a group under composition.

An example crucial to our discussion of the RSA encryption system is the group $\mathbb{Z}_n^*$, where n is a positive integer and $\mathbb{Z}_n$ denotes the ring of integers modulo n. By Corollary 1.17, if $\overline{a} \in \mathbb{Z}_n$, then there is a solution to the equation $\overline{a} \cdot \overline{x} = \overline{1}$ if $\gcd(a, n) = 1$. That is, $\overline{a}$ has a multiplicative inverse in $\mathbb{Z}_n$ if $\gcd(a, n) = 1$. The converse is also true; if $\gcd(a, n) = 1$, then we may write $1 = ax + by$ for some integers x, y, by Proposition 1.13. Then $\overline{1} = \overline{a} \cdot \overline{x}$, so $\overline{a}$ has a multiplicative inverse. We thus have the following description:

$$\mathbb{Z}_n^* = \{\overline{a} \in \mathbb{Z}_n : \gcd(a, n) = 1\} .$$

How many elements does this group have? For example, $\mathbb{Z}_4^* = \{\overline{1}, \overline{3}\}$ has two elements, $\mathbb{Z}_6^* = \{\overline{1}, \overline{5}\}$ also has two elements, and $\mathbb{Z}_{11}^* = \mathbb{Z}_{11} - \{\overline{0}\}$ has ten elements. In general, the number of elements of $\mathbb{Z}_n^*$ is equal to the number of integers a satisfying $1 \le a \le n$ and $\gcd(a, n) = 1$. This number is referred to as $\phi(n)$, and the function ϕ is called the *Euler phi function*. While there is a general formula for computing $\phi(n)$ from the prime factorization of n, we concern ourselves only with two cases. First, if p is a prime, then $\phi(p) = p - 1$. This is clear since all integers a with $1 \le a \le p$ except for p itself are relatively prime to p. The next case is stated as a lemma.

Lemma 8.16. *Let p and q be distinct primes. Then $\phi(pq) = (p - 1)(q - 1)$.*

Proof. Set $n = pq$. To count $\phi(n)$, we first count the number of integers between 1 and n and not relatively prime to n. If $1 \le a \le n$, then $\gcd(a, n) > 1$ only if p divides a or q divides a. The multiples of p between 1 and n are then

$$p, 2p, \ldots, (q - 1)p, qp = n,$$

so there are q multiples of p between 1 and n. The multiples of q in this range are

$$q, 2q, \ldots, (p - 1)q, pq = n,$$

so there are p multiples of q in this range. The only number on both lists is n; this follows from unique factorization. Therefore, there are $p+q-1$ integers between 1 and n that are not relatively prime to n. Since there are $n = pq$ numbers total in this range, we see that

$$\begin{aligned}\phi(n) &= pq - (p+q-1) = pq - p - q + 1 \\ &= p(q-1) - (q-1) = (p-1)(q-1),\end{aligned}$$

as claimed. □

To summarize, the group $\mathbb{Z}_n^*$ has $\phi(n)$ elements, and if $n = pq$ is the product of two distinct primes, then $\mathbb{Z}_{pq}^*$ has $(p-1)(q-1)$ elements. The significance of this result and its application to encryption rely on a basic result from abstract group theory, Lagrange's Theorem, which gives surprisingly strong information about the subgroups of a finite group, i.e., the group theoretic analogues of subspaces of a vector space.

8.1.1 Subgroups

Definition 8.17. Let G be a group. A nonempty subset H of G is said to be a subgroup of G if the operation on G restricts to an operation on H, and if H is a group with respect to this restricted operation.

For example, consider the group $\mathbb{Z}$ under addition and let H be the set of even integers. Addition restricts to an operation on H because the sum of two even integers is again even. Because 0 is even, as is the additive inverse of each even integer, H is a subgroup. Just as there is a theorem helping to determine when a subset of a vector space is a subspace (Lemma 4.9), there is a result that helps us determine when a subset of a group is a subgroup.

Lemma 8.18. *Let G be a group, and let H be a nonempty subset of G. Then H is a subgroup of G provided that the following two conditions hold: (i) if $a, b \in H$, then $ab \in H$, and (ii) if $a \in H$, then $a^{-1} \in H$.*

Proof. Suppose a subset H of a group G satisfies the two conditions in the statement. Condition *(i)* says that the operation on G restricts to an operation on H, so we have a binary operation on H. We need to verify for H the three axioms in the definition of a group. Associativity is clear; if $a, b, c \in H$, then $a, b, c \in G$, so $a(bc) = (ab)c$ since G is a group. Next, Condition *(ii)* ensures that every element of H has an inverse in H. So, the only thing remaining is to see that H has an identity. We do this by proving that if e is the identity of G, then $e \in H$. To see why this is true, first note that there is an element $a \in H$ because H is nonempty. By Condition *(ii)*, $a^{-1} \in H$. Then, by Condition *(i)*, $a \cdot a^{-1} \in H$. But $a \cdot a^{-1} = e$, so $e \in H$. □

Example 8.19. Let F be a field and $G = \mathrm{GL}_n(F)$ the group of nonsingular matrices with entries in F. Denote by $\mathrm{SL}_n(F)$ the subset consisting of matrices of determinant exactly equal to 1. Since, for every pair of matrices A and B, $\det(AB) = \det(A)\det(B)$ and, for A nonsingular, $\det(A^{-1}) = \det(A)^{-1}$, Lemma 8.18 shows that $\mathrm{SL}_n(F)$ is a subgroup of $\mathrm{GL}_n(F)$.

Example 8.20. Again let $G = \mathrm{GL}_n(\mathbb{R})$ and T be the subset of nonsingular triangular matrices. Despite the fact that the inverse of a nonsingular triangular matrix is again triangular, T is not closed under matrix–matrix multiplication. On the other hand, the subsets of nonsingular upper (resp. lower) triangular matrices are subgroups of G.

There is a particularly simple construction of subgroups. Let G be a group, take some $a \in G$, and use it to build a subgroup, call it H, containing a. By Condition *(i)* of Lemma 8.14, we see that every subgroup containing a has to contain $a \cdot a = a^2$, so $a^2 \in H$. Using the condition again, H has to contain $a^2 \cdot a = a^3$ and clearly H must also contain a^{-1}. Then by Condition *(i)*, it must contain $a^{-1} \cdot a^{-1} = a^{-2}$. An easy induction argument shows that $\{a^n : n \in \mathbb{Z}\} \subset H$. Conversely, an application of Lemma 8.14 shows that $\{a^n : n \in \mathbb{Z}\}$ is itself a subgroup of G, hence we make the following definition.

Definition 8.21. Let G be a group and $a \in G$, then the *cyclic subgroup* generated by a is the set

$$\langle a \rangle = \{a^n : n \in \mathbb{Z}\}.$$

Example 8.22. Let $S = \{z \in \mathbb{C} : |z| = 1\}$. Clearly $S \subset \mathbb{C}^*$ and an application of Lemma 8.14 shows that S is a subgroup of $\mathbb{C}^*$. Exercise 17 below shows that S is infinite and not a cyclic group. But S has many finite cyclic subgroups. For instance, let $\theta = e^{2\pi i/n}$ for n any positive integer. Then $\langle \theta \rangle = \{1, \theta, \theta^2, \ldots, \theta^{n-1}\}$ is one example.

Example 8.23. Let $G = \mathbb{Z}_8^*$. Then $G = \{\overline{1}, \overline{3}, \overline{5}, \overline{7}\}$. We calculate the cyclic subgroups of G. First, since $\overline{1}^n = \overline{1}$ for all n, we have $\langle \overline{1} \rangle = \{\overline{1}\}$ contains only the identity $\overline{1}$. Next, for $\overline{3}$, we see that

$$\overline{3}^1 = \overline{3},$$

$$\overline{3}^2 = \overline{9} = \overline{1},$$

$$\overline{3}^3 = \overline{27} = \overline{3},$$

$$\overline{3}^4 = \overline{81} = \overline{1}.$$

You may recognize a pattern. Since $\overline{3}^2 = \overline{1}$, the identity of G, if $n = 2m$ is even, then $\overline{3}^n = \overline{3}^{2m} = (\overline{3}^2)^m = \overline{1}^m = \overline{1}$. If $n = 2m + 1$ is odd, then

$$\overline{3}^n = \overline{3}^{2m+1} = \overline{3} \cdot \overline{3}^{2m} = \overline{3} \cdot \overline{1} = \overline{3}.$$

Therefore, the only powers of $\overline{3}$ are $\overline{1}$ and $\overline{3}$. Thus, $\langle \overline{3} \rangle = \{\overline{1}, \overline{3}\}$. If we do similar calculations for $\overline{5}$ and $\overline{7}$, we will see similar patterns, for $\overline{5}^2 = \overline{1}$ and $\overline{7}^2 = \overline{1}$. Thus, the only powers of $\overline{5}$ are $\overline{1}$ and $\overline{5}$, and the only powers of $\overline{7}$ are $\overline{1}$ and $\overline{7}$. Therefore $\{\overline{1}\}, \{\overline{1}, \overline{3},\}, \{\overline{1}, \overline{5},\}$, and $\{\overline{1}, \overline{7}\}$ are the cyclic subgroups of $\mathbb{Z}_8^*$

Example 8.24. Let $G = \mathbb{Z}_{10}^*$. Then $G = \{\overline{1}, \overline{3}, \overline{7}, \overline{9}\}$ has four elements. As with the previous example (and any group in fact), the cyclic subgroup generated by $\overline{1}$ is just $\{\overline{1}\}$. Next consider $\overline{3}$. We have

$$\overline{3}^1 = \overline{3},$$

$$\overline{3}^2 = \overline{9},$$

$$\overline{3}^3 = \overline{27} = \overline{7},$$

$$\overline{3}^4 = \overline{81} = 1.$$

We have therefore produced all four elements of G as powers of $\overline{3}$; therefore, $\langle\overline{3}\rangle = G$, i.e., G is a cyclic group. A similar calculation will show that $\langle\overline{7}\rangle = G$; this can also be seen by the fact that $\overline{7} = \overline{3}^{-1}$ and $\langle a\rangle = \langle a^{-1}\rangle$ for any group G and any $a \in G$. Finally, for $\overline{9}$, we have

$$\overline{9}^1 = \overline{9},$$

$$\overline{9}^2 = \overline{81} = \overline{1}.$$

As with the previous example, even powers of $\overline{9}$ are $\overline{1}$ and odd powers of $\overline{9}$ are $\overline{9}$, so $\langle\overline{9}\rangle = \{\overline{1}, \overline{9}\}$.

8.1.2 Lagrange's Theorem

We make some numerical observations from the previous two examples. The first thing to notice is that the number of elements in a given cyclic subgroup $\langle a\rangle$ turned out to be the same as the first positive number n for which a^n was the identity. Second, the number of elements in each cyclic subgroup was a divisor of the number of elements of the given group. Neither of these facts is a coincidence; they are general facts that we will now prove. The second is in fact a special case of Lagrange's Theorem. For a piece of notation, we will write $|X|$ for the number of elements in a set X. For a finite group G, the number $|G|$ is often called the *order* of G.

Example 8.25. Let p be a prime integer and $G = \mathrm{GL}_2(\mathbb{Z}_p)$. For a 2×2 matrix to lie in G, its first row can consist of any pair of residue classes mod p except for $(0, 0)$. Thus there are exactly $p^2 - 1$ possible first rows. Once the first row is determined, the second row can again consist of any pair of residue classes mod p except for a $\mathbb{Z}_p$-multiple of the first row. Thus $|G| = (p^2 - 1)(p^2 - p)$.

Lemma 8.26. *Let G be a group and $a \in G$. Set $n = \min\{m : m > 0, a^m = e\}$ if this number exists. Then $n = |\langle a\rangle|$, the number of elements in the cyclic subgroup generated by a. This number is called the order of the element a.*

Proof. We will prove the lemma by proving that $\langle a\rangle = \{a^r : 0 \le r < n\}$ and that these elements are all distinct. First, any element of $\langle a\rangle$ is of the form a^s for some integer s. By the Division Algorithm, we may write $s = qn + r$ with $0 \le r < n$. Then

$$a^s = a^{qn+r} = a^{qn}a^r = (a^n)^q(a^r) = a^r$$

since $a^n = e$, so $(a^n)^q = e$. Therefore, a^s can be written as a power a^r of a with $0 \le r < n$. This proves the first claim. For the second, suppose that $a^r = a^t$ with $0 \le r, t < n$. Suppose that $r \le t$. Then, by the laws of exponents, $e = a^t a^{-r} = a^{t-r}$. Since n is the smallest positive integer satisfying $a^m = e$, and since $0 \le t - r < n$, we must have $t - r = 0$. Thus, $t = r$. So, the elements $a^0, a^1, \ldots, a^{n-1}$ are all distinct. Since they form $\langle a\rangle$, we have proved that $|\langle a\rangle| = n$.

□

If G is infinite and $g \in G$ generates an infinite cyclic subgroup, then g is said to have infinite order. For instance, every nonzero integer has infinite order in $(\mathbb{Z}, +)$. On the other hand, -1 has order 2 in the infinite group $\mathbb{Q}^*$ and $e^{2\pi i/n}$ has order n in $\mathbb{C}^*$.

We now consider Lagrange's Theorem. To do so we need to look at a concept we have seen twice before, in the cases of vector spaces and rings. If H is a subgroup of a group G, and if $a \in G$, then the *right coset* of H determined by a is the set

$$Ha = \{ha : h \in H\}.$$

Right cosets are equivalence classes for the following relation: for $a, b \in G$, define $a \sim b$ if $ab^{-1} \in H$. In Exercise 7 below you will show that $\sim$ is an equivalence relation and that the equivalence class of a is the coset Ha. Therefore, the right cosets of H form a partition for the group G. If we write the group operation as multiplication, we will usually write Ha for the coset of a. Left cosets are defined analogously for the equivalence relation $a \sim b$ if $b^{-1}a \in H$ and are denoted

$$aH = \{ah : h \in H\}.$$

Theorem 8.27 (Lagrange's Theorem). *Let G be a finite group and let H be a subgroup of G. Then $|H|$ divides $|G|$.*

Proof. We prove this by showing that each right coset has $|H|$ elements (the identical argument, suitably modified, shows that each left coset also has $|H|$ elements). From this it will follow that $|G|$ is equal to the product of $|H|$ with the number of cosets, and this will prove the theorem. To that end, let $a \in G$, and we proceed to demonstrate that $|Ha| = |H|$. One way to prove that two sets have the same size is to produce a 1-1 onto function between them. This is done here by defining a function $f : H \to Ha$ by $f(h) = ha$. The function f is 1-1 since if $f(h) = f(k)$, then $ha = ka$. Multiplying both sides on the right by a^{-1} yields $h = k$, so f is 1-1. The function f is also onto since, if $x \in Ha$, then $x = ha$ for some $h \in H$, and so $x = f(h)$. Since f is then a 1-1 onto function from H to Ha, we have proven that $|H| = |Ha|$, as desired. □

In fact, we can be a little more detailed. We let $[G : H]$ be the number of right (and therefore also left) cosets of H in G. This number is often called the *index* of H in G. The proof of Lagrange's Theorem shows that all cosets have the same number of elements. Therefore, $|G|$ is equal to the product of the size of any one coset with the number of cosets. In other words, since $H = He$ is a coset, $|G| = |H| \cdot [G : H]$.

Lagrange's Theorem combined with Lemma 8.26 yields a result key to the RSA encryption system.

Corollary 8.28. *Let G be a finite group with $n = |G|$. If $a \in G$, then $a^n = e$.*

Proof. Let $m = |\langle a \rangle|$. By Lagrange's Theorem, m divides n, so $n = mt$ for some integer t. By Lemma 8.26, $a^m = e$. Therefore, $a^n = a^{mt} = (a^m)^t = e^t = e$, as desired. □

The following theorem due to Euler is a special case of this corollary.

Corollary 8.29 (Euler's Theorem). *Let n be a positive integer. If a is an integer with* $\gcd(a, n) = 1$, *then* $a^{\phi(n)} \equiv 1 \bmod n$.

Proof. If $\gcd(a, n) = 1$, then $\overline{a} \in \mathbb{Z}_n^*$, a group of order $\phi(n)$. The previous corollary tells us that $\overline{a}^{\phi(n)} = \overline{1}$. By definition of coset multiplication, $\overline{a}^{\phi(n)} = \overline{a^{\phi(n)}}$. The equation $\overline{a^{\phi(n)}} = \overline{1}$ is equivalent to the relation $a^{\phi(n)} \equiv 1 \bmod n$. □

Example 8.30. The special case of Euler's Theorem when $n = p$ is prime is the earlier result known as Fermat's Little Theorem. Noting that $\phi(p) = p - 1$, Euler's Theorem states that if the integer a is not divisible by p, then $a^{p-1} \equiv 1 \bmod p$. In fact $a^p \equiv a \bmod p$ for every integer

a. This follows from Fermat's Little Theorem by multiplying both sides if a is not a multiple of p, and the observation that a^p is a multiple of p if and only if the same s true for a.

Example 8.31. The case $n = pq$ for distinct primes p and q is important for RSA cryptography: If $\overline{a} \in \mathbb{Z}_n^*$, then $a^{(p-1)(q-1)} \equiv 1 \bmod n$.

Example 8.32. Let $G = \mathrm{GL}_2(\mathbb{Z}_p)$ and $H = \mathrm{SL}_2(\mathbb{Z}_p)$. To determine the order of H note, as for G, the first row of a matrix in H can be an arbitrary pair (a, b) of elements of $\mathbb{Z}_p$ except for $(0, 0)$ so there are $p^2 - 1$ possible first rows. Given one such row (a, b), the second row must consist of a solution in $\mathbb{Z}_p^2$ of $ax - by = 1$. If $a = 0$, then x can be arbitrary in $\mathbb{Z}_p$, but $b \neq 0$ so that $y = -b^{-1}$ is uniquely determined. If $a \neq 0$, then $x = a^{-1}(1 + by)$ is uniquely determined by y, which can be arbitrary. Either way, for each first row, there are exactly p second rows. Hence $|H| = p(p^2 - 1)$, which is a divisor of

$$\begin{aligned} |G| &= (p^2 - 1)(p^2 - p) \\ &= (p - 1)p(p^2 - 1). \end{aligned}$$

Exercises

1. Prove the cancellation law for a group: If G is a group and $a, b, c \in G$, then $b = c$ if and only if $ba = ca$ (if and only if $ab = ac$).
2. Let X be a set and $P(X)$ the set of all permutations of X under composition. Show that $P(X)$ is a group. Is the operation commutative? If so, prove it. If not, find a set X and $f, g \in P(X)$ with $f \circ g \neq g \circ f$.
3. Let G be a group. If $a \in G$ and n, m are positive integers, prove that

   (a) $a^n a^m = a^{n+m}$ and
   (b) $(a^n)^m = a^{nm}$.

   (Hint: Recall the inductive definition of a^n. For (a), use induction on m for (a). Then use (a) and induction on m for (b).)
4. Let G be a group. If $a, b \in G$ with $ab = ba$, show that $(ab)^n = a^n b^n$.
5. A group in which $ab = ba$ for every $a, b \in G$ is said to be abelian. Let G be a group in which every nonidentity element has order equal to 2 (i.e., $a^2 = e$ for every $a \in G$). Prove that G is abelian.
6. Let G be a group. If $a \in G$ and n and m positive integers, prove that $a^n a^m = a^{n+m}$, $(a^n)^m = a^{nm}$, and $(a^n)^{-1} = (a^{-1})^n$.
7. Let G be a group and let H be a subgroup of G. Prove that the relation $\sim$ on G given by $a \sim b$ if $ab^{-1} \in H$ is an equivalence relation.
8. With notation in the previous problem, if $a \in G$, show that the right coset Ha is the equivalence class of a.
9. Prove that every infinite cyclic group has infinitely many distinct subgroups.
10. Prove that a group is finite if and only if it has only finitely many distinct subgroups. One direction is clear. For the other consider the cyclic subgroups.
11. Prove that every group of prime order is cyclic.
12. Let $G = \{a, b, c, e\}$ be a group of order 4 with e the identity. Show that every such group is abelian. In particular prove that every such group is either cyclic or every nonidentity element has order 2
13. Prove that $\mathbb{Q}^*$ is not a cyclic group. (Hint: If $\frac{a}{b} \in \mathbb{Q}^*$ with $\gcd(a, b) = 1$ prove that $\frac{1}{b+1} \notin \langle \frac{a}{b} \rangle$.)
14. Write out the multiplication table for $\mathbb{Z}_{15}^*$.
15. Find the cyclic subgroups $\langle \overline{4} \rangle$ and $\langle \overline{7} \rangle$ of $\mathbb{Z}_{15}^*$.

16. Write out all the cosets in $\mathbb{Z}_{15}^*$ for the subgroup $\langle\overline{2}\rangle$.
17. Let G be a group and suppose that $a \in G$ has finite order equal to n. If $a^m = e$, prove that m is divisible by n.
18. Show that $S = \{z \in \mathbb{C} : |z| = 1\}$ is an infinite non-cyclic group. (Hint: For each real number $\alpha \in [0, 1), e^{2\pi i\alpha} \in S$.)

8.2 Cryptography and Group Theory

This section discusses one of the most common methods of encrypting highly sensitive data, namely the RSA encryption system. Cryptography is the subject of transmitting private data in a secure manner. If you make a purchase on the internet, you need to send to the merchant a credit card number. If somebody were to intercept the transmission of this information, your credit card number would be known to this person. Because of this, most Internet sites encrypt data identifying the user and his or her credit card numbers. By doing so, someone intercepting the transmission will see useless strings of digits instead of the user's identity and credit card number. If, however, the interceptor were to know how the merchant replaces the identifying information with other numbers, they would have a way of recovering the information. Because of this, merchants must use methods of encryption that are very difficult to "break."

8.2.1 The RSA Encryption System

RSA encryption was invented in 1978 by Ron Rivest, Adi Shamir, and Leonard Adleman. Despite the conceptual simplicity of the encryption/decryption algorithm, RSA has resisted virtually all attacks in the 36 years since its invention and is therefore in ubiquitous use for purposes like the exchange of passwords for identity verification. To describe the RSA system, one starts with the following data:

- Distinct prime numbers p and q, and
- an integer e relatively prime to $(p-1)(q-1)$.

From this data we will build an encryption system. We will restrict our attention to encrypting numbers. This is satisfactory, since any text message can be converted to numbers by replacing each letter with an appropriate number. Let M be an integer, considered to be a message we wish to encrypt. Set $n = pq$, the product of the two primes above. The integer e is the *encryption key*, in the sense that M is encrypted via the calculation $M^e \bmod n$, i.e., the remainder after dividing n into M^e. This remainder is our encrypted message.

For example, let

$$p = 3486784409,$$
$$q = 282429536483, \text{ and}$$
$$e = 19.$$

Then $n = pq = 984770904450021093547$. Also,

$$(p-1)(q-1) = 984770904164104772656$$

and one can check that $e = 19$ is relatively prime to this number.

Encryption of the message 12345 is effected by the calculation

$$12345^{19} \bmod n,$$

which comes out to be 123355218486796132288. Therefore, the encrypted message 12345 is transmitted as

$$123355218486796132288.$$

How does somebody receiving 123355218486796132288 know that this number represents 12345? First, by our assumption that e is relatively prime to $(p-1)(q-1)$, we know from Corollary 1.17 that e has a multiplicative inverse modulo $(p-1)(q-1)$. The integer d satisfying $ed \equiv 1 \bmod (p-1)(q-1)$, and chosen so that $0 < d < (p-1)(q-1)$, is the *decryption key*: If an encrypted number N is received, then one calculates $N^d \bmod n$; the result returns the original message. For example, from a Maple computation, we can see that

$$d = 207320190350337846875.$$

Thus, to recover the original message 12345, we compute

$$\begin{aligned}&123355218486796132288^{207320190350337846875} \bmod 984770904450021093547\\&= 12345.\end{aligned}$$

While this calculation looks formidable, Maple can do it virtually instantaneously. In fact, on an average personal computer, Maple can calculate $M^d \bmod n$ in a couple of seconds even if d and n are 400 digit numbers, so the calculations in the RSA system are easy to do even with very large numbers.

To summarize, the RSA encryption system starts with two prime numbers p and q, and an integer e satisfying $\gcd(e, (p-1)(q-1)) = 1$. One then calculates a positive integer d satisfying $ed \equiv 1 \bmod (p-1)(q-1)$. The integer message M is encrypted by replacing it with

$$N = M^e \bmod n.$$

To decrypt N, one sees that

$$M = N^d \bmod n.$$

The security of RSA cryptography is due to the practical difficulty of finding the prime factors of integers, a difficulty which increases with the magnitude of the prime factors. In practice RSA uses very large primes p and q (on the order of 100 digits each) resulting in encrypted messages with magnitudes of roughly 200 digits. This is one reason for the importance of the search for large prime numbers, and the related problem of *primality testing*, i.e., determining whether or not a given integer is prime.

The large sizes of encrypted messages use a significant amount of computer memory, so RSA is typically used for encryption of short messages, e.g., passwords for systems with many users. For this type of application, the encryption key e is public, i.e., made available to all users, hence the term "Public Key Cryptography." The decryption key, d, is possessed only by the intended recipient. The public uses the "public key" to lock up information, but only the intended recipient can unlock it with the "private key."

That the encryption/decryption method works is a direct consequence of Euler's Theorem applied in the group $G = \mathbb{Z}_n^*$, where $n = pq$ is the product of our two distinct prime numbers. As we have seen, $|G| = \phi(n) = (p-1)(q-1)$. Recall that the integers e and d satisfy $ed \equiv 1 \bmod \phi(n)$ so we may write $1 = ed + s\phi(n)$ for some integer s. The claim of the RSA system is that, for any message M, we have $(M^e)^d \equiv 1 \bmod n$. Written another way, it claims that $(\overline{M}^e)^d = \overline{M}$. Assuming that M is divisible by neither p nor q, we have $\overline{M} \in \mathbb{Z}_n^*$. Therefore

$$\begin{aligned}\overline{M} &= \overline{M}^{ed+s\phi(n)} = \overline{M}^{ed}\,\overline{M}^{s\phi(n)} = \overline{M}^{ed}\,(\overline{M}^{\phi(n)})^s \\ &= \overline{M}^{ed}\,(\overline{1})^s = \overline{M}^{ed}\end{aligned}$$

since $\overline{M}^{\phi(n)} = 1$ by Euler's Theorem. Thus, $(\overline{M}^e)^d = \overline{M}$, and so the decryption in RSA recovers the original message.

In the argument above, we assumed that M was divisible by neither p nor q in order to conclude that decryption would recover M. This is not a necessary assumption, but it makes the argument a little simpler. The case in which M is divisible by p or q is left to the exercises.

While the familiar Fundamental Theorem of Arithmetic asserts that every integer greater than 1 is either prime or a product of primes in a unique way up to the order of the primes, the security of the RSA system is based on the empirical fact that finding the prime factors is an inherently difficult problem; that is there is no known algorithm that efficiently finds those prime factors. Recall that the public aspect of RSA is that n and the encryption key e can be public knowledge. On the other hand, the factors p and q of n remain private. Were one to be able to factor n, then knowledge of p and q would enable the computation of $m = (p-1)(q-1)$ and hence the solution of the congruence $ex \equiv 1 \bmod m$ for the decryption key d.

Exercises

1. Let $n = 33$. Find an e, d pair and test the encryption/decryption procedure on several single digit integers.
2. Find an e, d pair for $n = 91$.
3. If an eavesdropper on an RSA encrypted communication could somehow find $m = (p-1)(q-1)$, then the encryption could be broken because it is easy to find the decoding integer from the publicly available e (Maple does it via the igcdex command). Show via elementary algebra how to determine p and q if you know m (and the public information $n = pq$). Give an example of how to do this by finding p and q from the values of n and m in the Maple worksheet Section-8.2-Exercise-3.mw.

   (Hint: You cannot do this with the factor command applied to the n in that worksheet because it is too large!)
4. Let $n = pq$ with p, q distinct primes. If a, b are integers, show that $a \equiv b \bmod n$ if and only if both $a \equiv b \bmod p$ and $a \equiv b \bmod q$.
5. Let (n, e, d) be RSA data and M an integer message to be encrypted. Verify that $M^{ed} \equiv M \bmod n$ if the integer M is divisible by either p or q.
6. Explain why the integer message M must be less than n.

8.2.2 Secure Signatures with RSA

One issue of data transmission is the ability to verify a person's identity. If you send a request to a bank to transfer money out of an account, the bank wants to know that you are the owner of the account.

If the request is made over the Internet, how can the bank verify your identity? The RSA encryption system gives a method for checking identities, which is one of the important features of the system.

Suppose that person A transmits data to person B, and that person B wants a method to check the identity of person A. To do this, both person A and B get sets of RSA data; person A has a modulus n_A and an encryption exponent e_A. These are publicly available. That person also has a decryption exponent d_A that remains private. Person B similarly has data n_B, e_B, and d_B. In addition, person A has a *signature,* a publicly available number S. To convince person B of his identity, person A first calculates $T = S^{d_A} \bmod n_A$ and then $R = T^{e_B} \bmod n_B$. He then transmits R to person B. Person B then decrypts R with her data, recovering $T = R^{d_B} \bmod n_B$. Finally, she encrypts T with person A's data, obtaining $T^{e_A} \bmod n_A = S$. By seeing that this result is the signature of person A, the identity has been validated.

For example, suppose that the data for person A is

$$\begin{aligned} n_A &= 2673157 \\ e_A &= 23 \\ d_A &= 2437607 \\ S &= 837361 \end{aligned}$$

and the data for person B is

$$\begin{aligned} n_B &= 721864639 \\ e_B &= 19823 \\ d_B &= 700322447. \end{aligned}$$

Person A then calculates

$$837361^{2437607} \bmod 2673157 = 1216606,$$

and

$$1216606^{19823} \bmod 721864639 = 241279367.$$

Person A then transmits 241279367 to person B. When person B receives this, he/she calculates

$$241279367^{700322447} \bmod 721864639 = 1216606,$$

and finally recovers S as $S = 1216606^{23} \bmod 2673157$.

To explain why this works, we denote by $\text{encrypt}_A(M)$ and $\text{decrypt}_A(M)$ the integers $M^{e_A} \bmod n_A$ and $M^{d_A} \bmod n_A$, respectively. We similarly have $\text{encrypt}_B(M)$ and $\text{decrypt}_B(M)$. The validity of the RSA system says that

$$\begin{aligned} \text{decrypt}_A(\text{encrypt}_A(M)) &= M, \\ \text{encrypt}_A(\text{decrypt}_A(M)) &= M. \end{aligned}$$

Similar equations hold for B. With this notation, person A calculates

$$R = \text{encrypt}_B(\text{decrypt}_A(S))$$

and then person B calculates

$$\text{encrypt}_A(\text{decrypt}_B(R)).$$

Therefore, person B will calculate

$$\begin{aligned}\text{encrypt}_A(\text{decrypt}_B(\text{encrypt}_B(\text{decrypt}_A(S)))) &= \text{encrypt}_A(\text{decrypt}_A(S)) \\ &= S\end{aligned}$$

because of the equations above. Therefore, person B does recover the signature of person A.

The reason that this method validates the identity of person A is because only person A can calculate $\text{decrypt}_A(S)$. If another person tries to claim he/she is person A, i.e., tries to substitute a number F in place of $\text{decrypt}_A(S)$, he/she will transmit $\text{encrypt}_B(F)$ to person B. Person B will then calculate

$$\text{encrypt}_A(\text{decrypt}_B(\text{encrypt}_B(F))) = \text{encrypt}_A(F).$$

However, in order to have $\text{encrypt}_A(F) = S$, we must have

$$\begin{aligned}\text{decrypt}_A(S) &= \text{decrypt}_A(\text{encrypt}_A(F)) \\ &= F,\end{aligned}$$

which means that this person has to have the correct decrypted number $\text{decrypt}_A(S)$; he/she cannot send any other number without person B realizing it is a fake number.

References

1. Herstein I (1975) Topics in algebra, 2nd edn. Wiley, Hoboken
2. Johnson B, Richman F (1997) Numbers and symmetry: an introduction to algebra. CRC, Boca Raton
3. Rotman J (1995) An introduction to the theory of groups (Graduate texts in mathematics), 4th edn. Springer, New York
4. Talbot J, Welsh D (2006) Complexity and cryptography: an introduction. Cambridge University Press, Cambridge

Chapter 9
The Structure of Groups

The application of group theory to cryptography discussed in the previous chapter utilized *abelian groups*, i.e., groups whose operation satisfies the commutative property. Nonabelian groups have also found application in many areas including cryptography, chemistry, physics, and even in interior and exterior decorating (wallpaper patterns and frieze patterns, respectively) as we'll see in the final chapter of this text. The present chapter develops some general structure theory of groups essential to these applications.

In modern mathematics the mappings between structures are as important objects of study as the structures themselves. Hence we study linear transformations of vector spaces and homomorphisms of rings. For groups, the relevant mappings are called group homomorphisms, which we proceed to investigate.

Definition 9.1. Let G and H be groups with operations $*_G$ and $*_H$. A function $f : G \to H$ is a group homomorphism provided $f(g_1 *_G g_2) = f(g_1) *_H f(g_2)$ for all $g_1, g_2 \in G$.

In other words, a group homomorphism respects the operations in the domain and codomain, (H, in this case). When the operations are clear from the context we adopt the usual convention of omitting them, i.e., the homomorphism satisfies $f(g_1 g_2) = f(g_1)f(g_2)$. Verification of the following properties is left to the exercises:

1. $f(e_G) = e_H$ and
2. $f(g^{-1}) = f(g)^{-1}$.

Example 9.2. Let $G = (\mathbb{Z}, +)$ and $H = (\mathbb{Z}_n, +)$. The rules for modular addition show that the function $\pi : G \to H$ given by $\pi(a) = \overline{a}$ is a group homomorphism.

Example 9.3. Let F be a field and $G = \mathrm{GL}_n(F)$. The multiplicative property of determinants shows that the determinant function $\det : G \to F^*$ is a group homomorphism.

Definition 9.4. An isomorphism $f : G \to H$ of groups is a homomorphism which is one-to-one and onto. An isomorphism $f : G \to G$ is called an automorphism of G.

Example 9.5. Let G be a group and $g \in G$. The function $c_g : G \to G$ given by $c_g(h) = ghg^{-1}$ is an automorphism. Indeed,

$$\begin{aligned} c_g(hk) &= ghkg^{-1} \\ &= gh(g^{-1}g)kg^{-1} \end{aligned}$$

D.R. Finston and P.J. Morandi, *Abstract Algebra: Structure and Application*,
Springer Undergraduate Texts in Mathematics and Technology, DOI 10.1007/978-3-319-04498-9_9

$$= (ghg^{-1})(gkg^{-1})$$
$$= c_g(h)c_g(k)$$

shows that c_g is a homomorphism. If $c_g(h) = c_g(k)$, then $ghg^{-1} = gkg^{-1}$. Premultiplying by g^{-1} and postmultiplying by g show that $h = k$ and therefore that c_g is one-to-one. Given arbitrary $a \in G$,

$$\begin{aligned} a &= (gg^{-1})a(gg^{-1}) \\ &= g(g^{-1}ag)g^{-1} \\ &= c_g(g^{-1}ag) \end{aligned}$$

so that c_g is onto.

If there is an isomorphism $f : G \to H$, the groups G and H are said to be isomorphic. This condition is denoted by $G \cong H$.

Example 9.6. Let $G = (\mathbb{R}, +)$ and $H = (\mathbb{R}^+, \cdot)$ where $\mathbb{R}^+ = \{\alpha \in \mathbb{R} : \alpha > 0\}$. Then the exponential function $\exp : \mathbb{R} \to \mathbb{R}^+$ is an isomorphism of groups.

Example 9.7. Let G and H be cyclic groups of the same order n. For instance, $n = p - 1$ for p prime, $G = (\mathbb{Z}_p^*, \cdot)$, and $H = \langle\theta\rangle \subset \mathbb{C}^*$ where $\theta = e^{\frac{2\pi i}{p-1}}$. Then G and H are isomorphic. To see this, let $G = \langle g\rangle$ and $H = \langle h\rangle$, so that the orders of g and h are precisely n. Define $f : G \to H$ by $f(g^i) = h^i$ for all i. Then f is clearly an onto homomorphism if it is well defined, i.e., if $g^i = g^j$ implies that $h^i = f(g^i) = f(g^j) = h^i$. But $g^i = g^j$ if and only if $g^{j-i} = e_G$, which in turn holds if and only if $j - i$ is a multiple of n the order of g. But this latter condition holds if and only if $h^i = h^i$. The same argument, with the roles of h and g reversed, shows that if $f(g^i) = f(g^j)$ then $g^i = g^j$, so that f is one-to-one.

Analogous to the kernel and image of a linear transformation of vector spaces, every group homomorphism $f : G \to H$ gives rise to a special subgroup of G and a special subgroup of H.

Definition 9.8. Let $f : G \to H$ be a group homomorphism.

1. The kernel of f is $\{g \in G : f(g) = e_H\}$ and is denoted by $\ker(f)$.
2. The image of f is $\{f(g) : g \in G\}$ and is denoted by $\mathrm{im}(f)$.

Example 9.9. Here are some natural examples of kernels and images.

1. The kernel of $\pi : (\mathbb{Z}, +) \to (\mathbb{Z}_n, +)$ is the subgroup $n\mathbb{Z}$ consisting of all integer multiples of n.
2. Let n be a positive integer. Multiplication by n is a homomorphism $\mu_n : \mathbb{Z} \to \mathbb{Z}$. The image of μ_n is $n\mathbb{Z}$.
3. The kernel of $\det : \mathrm{GL}_n(F) \to F^*$ is $\mathrm{SL}_n(F)$ and its image is F^*, since the diagonal matrix

$$\begin{pmatrix} \alpha & 0 & \cdots & 0 \\ 0 & 1 & \ddots & \vdots \\ \vdots & \ddots & \ddots & 0 \\ 0 & \cdots & 0 & 1 \end{pmatrix}$$

lies in $\mathrm{GL}_n(F)$ for every $\alpha \in F^*$.

Because a homomorphism preserves the group operation, its kernel has a special property: If $f : G \to H$ is a homomorphism and $k \in \ker(f)$, the computation

$$\begin{aligned} f(gkg^{-1}) &= f(g)f(k)f(g^{-1}) \\ &= f(g)e_H f(g)^{-1} \\ &= f(g)f(g)^{-1} = e_H \end{aligned}$$

shows that $gkg^{-1} \in \ker(f)$ for every $g \in G$. Subgroups with this property play a crucial role in the structure theory of groups and therefore merit their own name.

Definition 9.10. A subgroup N of the group G is said to be normal if $gng^{-1} \in N$ for every $n \in N$ and every $g \in G$.

In particular, the kernel of a group homomorphism is a normal subgroup. As will be seen below, the converse is also true, i.e., if N is a normal subgroup of the group G, then there is a group H and a homomorphism $f : G \to H$ whose kernel is precisely N.

Example 9.11. $\mathrm{SL}_n(\mathbb{R})$ is a normal subgroup of $\mathrm{GL}_n(\mathbb{R})$.

Example 9.12. If G is an abelian group, then every subgroup of G is normal. Clearly, in this case $gng^{-1} = gg^{-1}n = n$ for every pair of elements $g, n \in G$.

Normality of a subgroup can be expressed in terms of its cosets. Recall that for H a subgroup of the group G and $g \in G$, the left coset of H determined by g is $\{gh : h = H\}$ and is denoted by gH. Similarly, the right coset of H determined by g is $Hg = \{hg : h \in H\}$ and, if G is finite, then the proof of Lagrange's Theorem shows that the index $[G : H]$ of H in G, i.e., the common number of right cosets and left cosets is equal to $\frac{|G|}{|H|}$. The condition that H is normal is that $gH = Hg$ for every $g \in G$, i.e., that every left coset is also a right coset. This is exploited in the following useful proposition.

Proposition 9.13. *Let G be a group and let H be a subgroup of index 2. Then H is a normal subgroup of G.*

Proof. Choose an element g in G but not in H, so that H and gH are the two left cosets of H. Since G is the disjoint union of the left cosets of H, it must be the case that gH is the set theoretic complement of H in G. But G is also the disjoint union of the right cosets of H, so that Hg is also the set theoretic complement of H, i.e., $gH = Hg$ for every $g \notin H$. But certainly $hH = Hh = H$ for every $h \in H$ as well. □

Example 9.14. Let X be the three element set $\{1, 2, 3\}$ and let $P(X)$ be its group of permutations. The following notation is useful. Write

$$\begin{pmatrix} 1\,2\,3 \\ 3\,1\,2 \end{pmatrix}$$

for the permutation sending $1 \mapsto 3, 2 \mapsto 1$, and $3 \mapsto 2$. Then $P(X)$ has the 6 elements:

$$\begin{pmatrix} 1\,2\,3 \\ 1\,2\,3 \end{pmatrix}, \begin{pmatrix} 1\,2\,3 \\ 2\,3\,1 \end{pmatrix}, \begin{pmatrix} 1\,2\,3 \\ 3\,1\,2 \end{pmatrix},$$

$$\begin{pmatrix} 1\,2\,3 \\ 1\,3\,2 \end{pmatrix}, \begin{pmatrix} 1\,2\,3 \\ 3\,2\,1 \end{pmatrix}, \begin{pmatrix} 1\,2\,3 \\ 2\,1\,3 \end{pmatrix}.$$

One checks easily that the first three permutations comprise a subgroup of index 2, hence a normal subgroup, of $P(X)$. Each of the permutations in the second row has order 2 and therefore, together with the identity permutation, forms a subgroup of index three. None of these order two subgroups is normal however. For instance,

$$\begin{pmatrix} 1\,2\,3 \\ 1\,3\,2 \end{pmatrix} \circ \begin{pmatrix} 1\,2\,3 \\ 3\,2\,1 \end{pmatrix} \circ \begin{pmatrix} 1\,2\,3 \\ 1\,3\,2 \end{pmatrix}^{-1} = \begin{pmatrix} 1\,2\,3 \\ 2\,1\,3 \end{pmatrix},$$

which does not lie in the subgroup

$$\left\{ \begin{pmatrix} 1\,2\,3 \\ 1\,2\,3 \end{pmatrix}, \begin{pmatrix} 1\,2\,3 \\ 3\,2\,1 \end{pmatrix} \right\}.$$

9.1 Direct Products

We consider a method to construct bigger groups from smaller groups or, to think of it another way, to decompose a group into simpler subgroups. The direct product is a natural way to define a group structure on the set theoretic Cartesian product of two groups. The semidirect product generalizes this construction and is necessary to our investigation of symmetry; its development is left to Chap. 10.

Given two groups G_1 and G_2, there is a natural "componentwise" operation on the Cartesian product of the sets $G_1 \times G_2$:

$$(g_1, g_2) * (h_1 h_2) = (g_1 *_1 h_1, g_2 *_2 h_2)$$

where $g_1, h_1 \in G_1, g_2, h_2 \in G_2$ and $*_1, *_2$ are the operations in the groups G_1 and G_2 respectively. The proof of the following theorem is straightforward and is left as Exercise 8.

Theorem 9.15. *Let G_1 and G_2 be groups. The Cartesian product of the sets $G_1 \times G_2$ with the componentwise operation $(G_1 \times G_2, *)$ has the structure of a group.*

Definition 9.16. The group $(G_1 \times G_2, *)$ is called the direct product of G_1 and G_2.

Example 9.17. The simplest case of a nontrivial direct product is the group $\mathbb{Z}_2 \times \mathbb{Z}_2$, known as the Klein 4-group (after the great German mathematician Felix Klein, and the number of its elements). In fact $\mathbb{Z}_2 \times \mathbb{Z}_2$ as a set is exactly the $\mathbb{Z}_2$ vector space $\mathbb{Z}_2^2$ with the group operation being vector addition.

Example 9.18. Let p and q be distinct prime integers and C_p, C_q the cyclic groups of orders p and q written multiplicatively. Then $C_p \times C_q$ is cyclic of order pq. Indeed, let a be a generator of C_p and b a generator of C_q, and consider the element $(a, b) \in C_p \times C_q$. By Lagrange's Theorem its order can only be $1, p, q$, or pq. Since (a, b) is not the identity element of $C_p \times C_q$, its order is not 1. From $(a,b)^p = (a^p, b^p) = (1, b^p)$ and the fact that q does not divide p, we find that the order is not p. Reversing the roles of p and q we see that the order is not q. Therefore $C_p \times C_q$, which has pq elements, has an element of order pq, i.e., is cyclic.

It is sometimes possible to study the structure of a group by recognizing it as a direct product of certain of its subgroups. For instance, turning the previous example around, we can see that the cyclic group C_{pq} decomposes as the direct product of two cyclic subgroups of orders p and q respectively. Suppose that $a \in C_{pq}$ has order pq. Then $a^q := h$ must have order p and therefore generates a subgroup H of order p. Symmetrically, $a^p := k$ generates a subgroup K of order q. By Lagrange's Theorem, $H \cap K$ consists only of the identity element e of C_{pq}.

Moreover, the pq products $h^i k^j$ for $0 \le i \le p$, and $0 \le j \le q$ are necessarily distinct: $h^i k^j = h^{i'} k^{j'}$ implies that $h^i h^{-i'} = k^{-j'} k^{j'} \in H \cap K = \{e\}$. But $h^i h^{-i'} = k^{-j'} k^{j'} = e$ clearly implies that $h^i = h^{i'}$ and $k^j = k^{j'}$. In particular, the function

$$\mu : H \times K \to C_{pq}$$

$$\mu : (h^i, k^j) \mapsto h^i k^j$$

is well defined and one-to-one. Because both $H \times K$ and C_{pq} have pq elements, the Pigeonhole Principle implies that μ is also onto. It is straightforward to verify that

$$\mu((h^i, k^j)(h^{i'}, k^{j'})) = \mu((h^i, k^j)\mu((h^{i'}, k^{j'})),$$

i.e., μ respects the group operations on both sides, effecting the desired decomposition.

The next theorem formalizes the previous paragraph as a criterion to detect direct products.

Theorem 9.19. *Let G be a group with subgroups H and K. Then $G \cong H \times K$ if*

1. *$G = HK$ (i.e., every element $g \in G$ can be written as $g = hk$ with $h \in H$ and $k \in K$),*
2. *both H and K are normal subgroups of G, and*
3. *$H \cap K = \{e\}$.*

In this case the representation $g = hk$ is unique.

Proof. Imitating the previous example, define a function $f : H \times K \to G$ by $f(h,k) = hk$. By 1., f is surjective. If $h_1 k_1 = h_2 k_2$, then $h_2^{-1} h_1 = k_2 k_1^{-1} \in H \cap K = \{e\}$ by 3. But this forces $h_1 = h_2$ and $k_1 = k_2$ so that f is also one-to-one. It remains to show that f is a group homomorphism. First observe that for every $h \in H$ and $k \in K$,

$$hkh^{-1}k^{-1} = (hkh^{-1})k^{-1} \in K$$

because K is a normal subgroup, but also

$$hkh^{-1}k^{-1} = h(kh^{-1}k^{-1}) \in H$$

because H is a normal subgroup. Thus $hkh^{-1}k^{-1} = \{e\}$, i.e., $hk = kh$. Finally,

$$\begin{aligned} f((h_1, k_1)(h_2, k_2)) &= f(h_1 h_2, k_1 k_2) = h_1 h_2 k_1 k_2 \\ &= h_1 k_1 h_2 k_2 = f((h_1, k_1)) f(h_2, k_2)). \end{aligned}$$

□

See Exercise 8 for the converse to this theorem.

Example 9.20. Let $G = G_1 \times G_2$ and let $\pi_1 : G \to G_1$ be the projection to the first coordinate:

$$\pi_1(g_1, g_2) = g_1.$$

It is simple to check that π_1 is a homomorphism and that $\text{im}(\pi_1) = G_1$. With e_1 denoting the identity of G_1, the kernel of π_1 is clearly $\{(e_1, g_2) : g_2 \in G_2\}$. Denote $\ker(\pi_1)$ by $\widetilde{G_2}$, which is a subgroup of G that is evidently isomorphic to G_2. Consider now the set $G/\widetilde{G_2}$ of left cosets of $\widetilde{G_2}$. Recall that for $g \in G$, the left coset $g\widetilde{G_2}$ is simply the subset of G consisting of $\{gh : h \in \widetilde{G_2}\}$ and for $h \in \widetilde{G_2}$, the coset $h\widetilde{G_2} = \widetilde{G_2}$. Writing elements $g \in G$ as ordered pairs $g = (g_1, g_2)$ with $g_1 \in G_1$ and $g_2 \in G_2$,

$$\begin{aligned} G/\widetilde{G_2} &= \{(g_1, g_2)\widetilde{G_2} : (g_1, g_2) \in G\} \\ &= \{(g_1, e_2)(e_1, g_2)\widetilde{G_2} : (g_1, g_2) \in G\} \\ &= \{(g_1, e_2)\widetilde{G_2} : g_1 \in G_1\}. \end{aligned}$$

We can identify $G/\widetilde{G_2}$ as a set with $G_1 = \operatorname{im}(\pi_1)$. In fact $G/\widetilde{G_2}$ has a natural structure of a group which is isomorphic (as a group) to G_1 and, symmetrically, $G/\widetilde{G_1} \cong G_2$. Indeed, setting

$$(g_1, g_2)\widetilde{G_2}(g_1', g_2')\widetilde{G_2} = (g_1g_1', e_2)\widetilde{G_2},$$

the assertion is clear if this is a well-defined operation. But $(g_1, g_2)\widetilde{G_2} = (h_1, h_2)\widetilde{G_2}$ if and only if $(h_1^{-1}g_1, h_2^{-1}g_2) \in \widetilde{G_2}$, i.e., if and only if $h_1 = g_1$. That the operation is well defined now follows easily.

9.2 Normal Subgroups, Quotient Groups, and Homomorphisms

Embedded in the previous example are special cases of two important theorems. The first one realizes every normal subgroup as the kernel of some group homomorphism. The second one demonstrates the special role of this homomorphism. First some terminology.

Definition 9.21. A group homomorphism is said to be injective if it is a one-to-one function and surjective if it is onto.

Theorem 9.22. *Let G be a group and N a normal subgroup. Then G/N has a natural structure of a group and the function $q_N : G \to G/N$ given by $q_N(g) = gN$ is a surjective homomorphism with kernel N.*

Before proceeding to the proof of this theorem, the word "natural" in the statement deserves some explanation. The group structure on G/N is inherited from that of G in the following sense: given two left cosets gN and hN, we naturally enough set $gNhN$ equal to the left coset $(gh)N$ written ghN. The issue to contend with is that for every $n \in N$ we have $gN = gnN$ because

$$(gn)^{-1}gn = n^{-1}g^{-1}gn = e \in N.$$

Reminiscent of our discussion of the ring structure on R/I, where I is an ideal of the ring R (in particular Lemma 5.16), the left cosets are not uniquely represented in the form gN, but the group operation is defined in terms of this representation. As in the example above, the main issue is that normality of the subgroup N guarantees that the natural group operation is well defined.

Proof. Setting $gNhN = ghN$, it is necessary to verify that if $g_1N = g_2N$ and $h_1N = h_2N$ then $g_1h_1N = g_2h_2N$. In other words, if $g_2^{-1}g_1 \in N$ and $h_2^{-1}h_1 \in N$, then

$$(g_2h_2)^{-1}g_1h_1 \in N.$$

Calculating

$$\begin{aligned}(g_2h_2)^{-1}g_1h_1 &= h_2^{-1}g_2{}^{-1}g_1h_1 \\ &= h_2^{-1}(h_1h_1^{-1})g_2{}^{-1}g_1h_1 \\ &= (h_2^{-1}h_1)h_1^{-1}(g_2{}^{-1}g_1)h_1\end{aligned}$$

we have the factors $h_2^{-1}h_1 \in N$ by hypothesis and $h_1^{-1}(g_2{}^{-1}g_1)h_1 \in N$ because $g_2^{-1}g_1 \in N$ and N is normal. The product in G/N is therefore independent of how the cosets are represented. Associativity of the product is clear, as are the facts that eN is the identity for this operation and $(gN)^{-1} = g^{-1}N$. Since the group structure on G/N is well defined, it is now clear that $q_N : G \to G/N$ given by $q_N(g) = gN$ is a surjective homomorphism. □

Definition 9.23. The group G/N is called the *quotient group* of G by the normal subgroup N.

Recall that if V is a vector space over the field F, then V is an abelian group under addition, and the zero vector is the identity element for this structure. Restricting to their abelian group structures, it is clear from the definition that a linear transformation of vector spaces $T : V \to W$ is an abelian group homomorphism. The next lemma is the generalization to arbitrary group homomorphisms of the fact that T is one-to-one if and only if its kernel consists only of the zero vector.

Lemma 9.24. *Let $f : G \to H$ be a group homomorphism. Then f is injective if and only if* $\ker f = \{e_G\}$.

Proof. Since $f(g_1) = f(g_2)$ if and only if $f(g_1)f(g_2)^{-1} = e_H$, and $f(g_1)f(g_2)^{-1} = f(g_1g_2{}^{-1})$, it follows that $f(g_1) = f(g_2)$ if and only iff $g_1g_2{}^{-1} \in \ker f$. Thus $\ker f = \{e_G\}$ if and only if $f(g_1) = f(g_2)$ implies that $g_1g_2{}^{-1} = e_G$, i.e., that $g_1 = g_2$. □

The next fundamental theorem shows that every group homomorphism "factors" into a surjective homomorphism followed by an injective homomorphism.

Theorem 9.25. *Let $f : G \to H$ be a group homomorphism with kernel K. Then there is a unique injective homomorphism $\tilde{f} : G/K \to H$ satisfying $f = \tilde{f} \circ q_K$. In particular, $G/K \cong \mathrm{im}(f)$.*

Proof. Define $\tilde{f}(gK) = f(g)$, noting that this is precisely the required condition $f = \tilde{f} \circ q_K$. In particular, this condition uniquely determines $\tilde{f}$ as a function of sets provided, as above, that $\tilde{f}$ is well defined. Suppose that $g_1K = g_2K$ so that $g_2^{-1}g_1 \in K = \ker f$. Then $e_H = f(g_2^{-1}g_1) = f(g_2)^{-1}f(g_1)$ so that

$$\tilde{f}(g_1K) = f(g_1) = f(g_2) = \tilde{f}(g_2K).$$

From the group operation in G/K and the definition of $\tilde{f}$, it follows for all $g_1, g_2 \in G$ that

$$\begin{aligned}\tilde{f}(g_1K)\tilde{f}(g_2K) &= f(g_1)f(g_2) = f(g_1g_2) \\ &= \tilde{f}(g_1g_2K) = \tilde{f}(g_1Kg_2K),\end{aligned}$$

i.e., that $\tilde{f}$ is a homomorphism.

To see that $\tilde{f}$ is injective, we determine the kernel and use the lemma above. A coset $gK \in \ker \tilde{f}$ if and only if $e_H = \tilde{f}(gK) = f(g)$, i.e., if and only if $g \in K$, so that $gK = eK = e_{G/K}$. □

Example 9.26. Let $G = (\mathbb{R}, +)$ and $K = (\mathbb{Z}, +)$. The rules of exponentiation show that the function $\varphi : G \to \mathbb{C}^*$ given by $\varphi(\alpha) = e^{2\pi i}$ is a group homomorphism. The kernel of φ is clearly equal to K, so Theorem 9.25 yields that G/K is isomorphic to its image as a subgroup of $\mathbb{C}^*$. The image of φ consists of

$$\{e^{2\pi i\theta} : \theta \in \mathbb{R}\} = \{\cos\theta + i\sin\theta : \theta \in \mathbb{R}\},$$

which is identified with the unit circle S^1 in $\mathbb{R}^2$ via

$$\cos\theta + i\sin\theta \mapsto (\cos\theta, \sin\theta).$$

We conclude that $\mathbb{R}/\mathbb{Z}$, with its structure as a group under addition, is isomorphic to S^1 as a group under multiplication.

Example 9.27. Let $G = \langle a \rangle$ be a cyclic group with $|G| = \infty$. Define a function $f : \mathbb{Z} \to G$ by $f(n) = a^n$. The laws of exponents show that f is a group homomorphism; it is onto by the definition of $G = \{a^n : n \in \mathbb{Z}\}$. The kernel of f is

$$\ker(f) = \{n \in \mathbb{Z} : f(n) = e\} = \{n \in \mathbb{Z} : a^n = e\}.$$

By Lemma 8.26, since G has infinite order, there is no nonzero n with $a^n = e$. Therefore, $\ker(f) = \{0\}$. Consequently, f is 1-1 and onto, which shows $G \cong \mathbb{Z}$.

Example 9.28. Let $G = \langle a \rangle$ be a cyclic group with $|G| = m$. As in the previous example, define $f : Z \to G$ by $f(n) = a^n$. The argument in the previous example shows that f is a group homomorphism and f is onto. However, by Lemma 8.26, $\ker(f) = \{mr : r \in \mathbb{Z}\}$ and by Theorem 9.25, G is isomorphic to the quotient group $\mathbb{Z}/\ker(f)$. But, since $\ker(f)$ is the set of multiples of m, this quotient group is precisely $\mathbb{Z}_m$. Thus, $G \cong \mathbb{Z}_m$.

As an application of Theorems 9.19 and 9.25, we have the following refinement of Exercise 12 of Sect. 8.1, to classify groups of order four.

Proposition 9.29. *Let G be a group of order 4. Then either $G \cong \mathbb{Z}_4$ or $G \cong \mathbb{Z}_2 \times \mathbb{Z}_2$.*

Proof. Suppose first that G has an element a of order 4. Then $G = \langle a \rangle$ is cyclic. The previous example then yields $G \cong \mathbb{Z}_4$. Next, suppose that G has no element of order 4. By Lagrange's theorem, the order of each element of G divides the order of G, which is 4. Thus, since G has no element of order 4, the order of each element must be either 1 or 2. Because the only element of a group of order 1 is the identity, we see that $G = \{e, a, b, c\}$ with a, b, c each having order 2. Let $H = \langle a \rangle$ and $K = \langle b \rangle$. Then H and K both have order 2. By Lagrange's theorem, both have index 2 in G, and so each is a normal subgroup of G by Proposition 9.13. Because $H = \{e, a\}$ and $K = \{e, b\}$, we see that $H \cap K = \{e\}$. Furthermore, $HK = \{e, a, b, ab\}$. We must have $ab = c$, and so $HK = G$. Thus, by Theorem 9.19, G is the direct product of H and K. By the previous example, both H and K are isomorphic to $\mathbb{Z}_2$. Thus, $G \cong \mathbb{Z}_2 \times \mathbb{Z}_2$. □

Quite generally, but with a bit more of advanced group theory, it can be shown that if p is any prime number and G is a group of order p^2, then

1. G is abelian, and
2. either $G \cong \mathbb{Z}_{p^2}$ or $G \cong \mathbb{Z}_p \times \mathbb{Z}_p$.

A word of caution is in order. Several arguments have used the product of two subgroups H and K of a group G and this construction will resurface in Chap. 10. The product of the subgroups H and

K is the set $HK = \{hk : h \in H, k \in K\}$ which is not in general a subgroup of G. A condition that guarantees that HK is a subgroup is given in Exercise 11.

For an example where HK fails to be a subgroup, take G to be the group of permutations of the three element set $\{1, 2, 3\}$ as in Example 9.14, with H and K any pair of distinct two element subgroups. Then HK is not a subgroup of G. For instance, take

$$H = \left\{ \begin{pmatrix} 1\,2\,3 \\ 1\,3\,2 \end{pmatrix}, \begin{pmatrix} 1\,2\,3 \\ 1\,2\,3 \end{pmatrix} \right\}, \; K = \left\{ \begin{pmatrix} 1\,2\,3 \\ 3\,2\,1 \end{pmatrix}, \begin{pmatrix} 1\,2\,3 \\ 1\,2\,3 \end{pmatrix} \right\}.$$

Then

$$\begin{pmatrix} 1\,2\,3 \\ 1\,3\,2 \end{pmatrix} \circ \begin{pmatrix} 1\,2\,3 \\ 3\,2\,1 \end{pmatrix} = \begin{pmatrix} 1\,2\,3 \\ 2\,3\,1 \end{pmatrix}$$

so that $HK =$

$$\left\{ \begin{pmatrix} 1\,2\,3 \\ 1\,2\,3 \end{pmatrix}, \begin{pmatrix} 1\,2\,3 \\ 1\,3\,2 \end{pmatrix}, \begin{pmatrix} 1\,2\,3 \\ 3\,2\,1 \end{pmatrix}, \begin{pmatrix} 1\,2\,3 \\ 2\,3\,1 \end{pmatrix} \right\}.$$

Lagrange's Theorem, or a calculation to show that HK is not closed under composition and inverses, demonstrates that this subset of G is not a subgroup. Another example is given in Exercise 11. In Exercise 12, a method is given to calculate $|HK|$ for G finite, whether or not HK is a subgroup.

Exercises

1. Prove for groups G_1 and G_2 :

   (a) that $G_1 \times G_2$ with the componentwise operation has the structure of a group,
   (b) that both $\widetilde{G_1} := G_1 \times \{e_2\}$ and $\widetilde{G_2} := \{e_1\} \times G_2$ are normal subgroups of $G_1 \times G_2$ whose intersection is the identity element of $G_1 \times G_2$,
   (c) that elements of $\widetilde{G_1}$ and $\widetilde{G_2}$ commute, i.e., for all $a \in \widetilde{G_1}$ and $b \in \widetilde{G_2}$ we have $ab = ba$.

2. Prove that the direct product of abelian groups is abelian but the direct product of two cyclic groups need not be cyclic.
3. Generalize Example 9.18 to give a criterion for the direct product of two cyclic groups to be cyclic.
4. For a group homomorphism $f : G \to H$ prove that:

   (a) $f(e_G) = e_H$;
   (b) $f(g^{-1}) = f(g)^{-1}$,
   (c) $\mathrm{im}(f)$ is a subgroup of H, and
   (d) $\ker(f)$ is a normal subgroup of G.

5. Let G be a group. Prove that G is abelian if and only if the function $s : G \to G$ given by $s(a) = a^2$ is a homomorphism.
6. Let G be a group and let $\mathrm{Aut}(G)$ be the set of all group automorphisms of G (Definition 9.4). Prove that $\mathrm{Aut}(G)$ is a group under the operation of function composition.
7. Let G be a group and $g \in G$. Referring to Example 9.5, prove that the function $c : G \to \mathrm{Aut}(G)$ given by $c(g) = c_g$ is a group homomorphism.
8. Using Theorem 9.25 prove, for every field F, that $\mathrm{GL}_n(F)/\mathrm{SL}_n(F) \cong F^*$.
9. Let G be the set of $n \times n$ permutation matrices, i.e., the subset of $\mathrm{M}_n(\mathbb{Z})$ consisting of those matrices with precisely one 1 in each row and in each column. Prove that

(a) G is a group under matrix multiplication,
(b) every element of G has determinant 1 or -1,
(c) the matrices of determinant equal to 1 comprise a normal subgroup N, and
(d) G/N is cyclic of order 2.

10. Let $p > 2$ be a prime number and let G be a group of order $2p$.

(a) Prove that G has an element (hence subgroup) of order p.
(b) Show that if H and K are subgroups of order p then $H \cap K = \{e\}$.
(c) Prove that G has a unique subgroup of order p and that it is normal.
(d) Prove that G has a subgroup of order 2. Note that Example 9.14 shows that this subgroup need not be normal.

11. Given a group G and subgroups H and K, the set HK is the subset $\{hk : h \in H, k \in K\}$ of G.

(a) With

$$G = \underset{n}{\mathrm{GL}}(\mathbb{R}),\ H = \left\{ \begin{pmatrix} 1 & a \\ 0 & 1 \end{pmatrix} : a \in \mathbb{R} \right\},\ K = \left\{ \begin{pmatrix} 1 & 0 \\ a & 1 \end{pmatrix} : a \in \mathbb{R} \right\},$$

show that HK is not a subgroup of G.
(b) Imitating the proof of Theorem 9.19, prove that HK is a subgroup of G if either H or K is a normal subgroup of G.

12. Given a group G and not necessarily normal subgroups H and K, it is clear that $h_1 k_1 = h_2 k_2$ implies $k_1 k_2^{-1} = h_1^{-1} h_2 \in H \cap K$. Conversely, for each $g \in H \cap K$ we have $hk = (hg)(g^{-1}k)$. Use these facts to prove that if G is finite, then $|HK| = \frac{|H||K|}{|H \cap K|}$.

References

1. Herstein I (1975) Topics in algebra, 2nd edn. Wiley, Hoboken
2. Rotman J (1995) An introduction to the theory of groups (Graduate texts in mathematics), 4th edn. Springer, New York

Chapter 10
Symmetry

In this chapter we explore another connection between algebra and geometry. One of the main issues studied in plane geometry is congruence; roughly, two geometric figures are said to be congruent if one can be moved to coincide exactly with the other. We will be more precise below in our description of congruence, and investigating this notion will lead us to new examples of groups. The culmination of this discussion is the mathematical classification of frieze patterns and wallpaper patterns based on the structure of the groups that arise.

10.1 Isometries

The following two triangles:

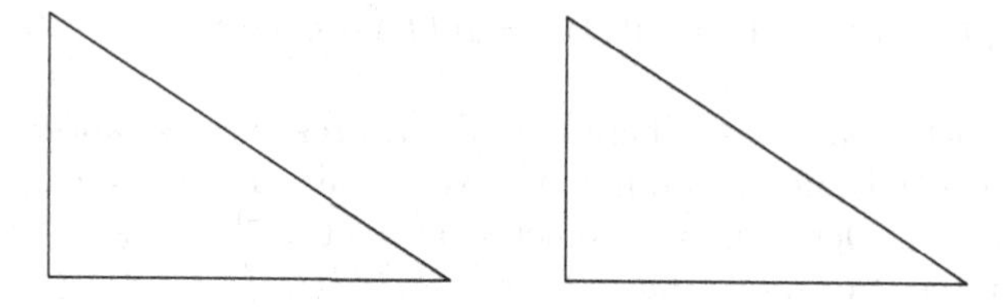

are congruent since the first can be moved to coincide with the second. In contrast, the next pair consists of non-congruent triangles:

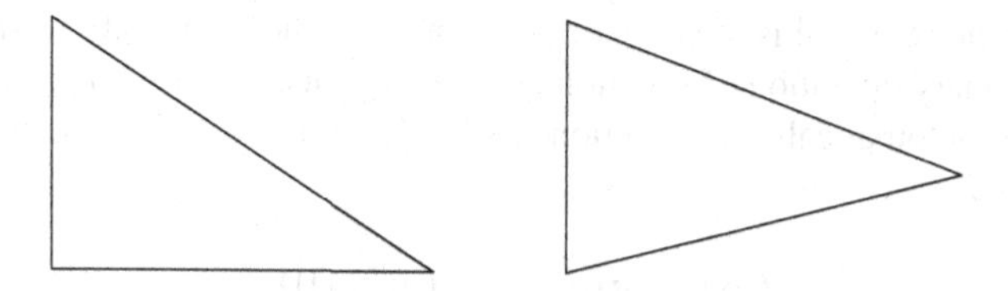

D.R. Finston and P.J. Morandi, *Abstract Algebra: Structure and Application*,
Springer Undergraduate Texts in Mathematics and Technology, DOI 10.1007/978-3-319-04498-9_10

since, in order to move the first to coincide with the second, one would also have to reshape the triangle. More precisely, they are not congruent because the distances between the vertices of the first triangle are not the same as the distances between the vertices of the second. The idea of distance enables a rigorous notion of congruence.

Definition 10.1. An isometry of the plane is a 1-1 and onto map of the plane to itself that preserves distances. That is, an isometry $f : \mathbb{R}^2 \to \mathbb{R}^2$ is a 1-1 onto function such that, for every pair P, Q of points in $\mathbb{R}^2$, the distance between P and Q is equal to the distance between $f(P)$ and $f(Q)$.

Two geometric figures are then *congruent* if there is an isometry that maps one figure exactly to the other. By representing points as ordered pairs of real numbers, we can give algebraic formulas for the distance between points. First, let $\|P\|$ be the distance from P to the origin $O = (0,0)$; this is the length of the line segment OP. The distance between $P = (a,b)$ and $Q = (c,d)$ is then given by the distance formula

$$\|P - Q\| = \sqrt{(a-c)^2 + (b-d)^2}.$$

The collection of all isometries of the plane $\mathbb{R}^2$ presents us with another important example of a group. Recalling the axioms defining this algebraic structure, the operation at hand is composition of isometries as functions from $\mathbb{R}^2$ to $\mathbb{R}^2$. Note that since an isometry is a 1-1 and onto function, it has an inverse function.

Lemma 10.2.

1. *The composition of two isometries of the plane is again an isometry.*
2. *The inverse of an isometry is an isometry.*

Proof. To prove (1), let f and g be isometries of the plane and let P, Q be points. Then

$$\|(f \circ g)(P) - (f \circ g)(Q)\| = \|g(P) - g(Q)\| = \|P - Q\|$$

since f and g are each isometries. Therefore, $f \circ g$ preserves distance. Moreover, $f \circ g$ is the composition of two 1-1 and onto functions, so it is also 1-1 and onto. Therefore, $f \circ g$ is an isometry. For part (2), let f be an isometry, and let f^{-1} be its inverse function. Let P and Q be points. We need to prove that $\left\|f^{-1}(P) - f^{-1}(Q)\right\| = \|P - Q\|$. Let $P' = f^{-1}(P)$ and $Q' = f^{-1}(Q)$. Then $P = f(P')$ and $Q = f(Q')$ by definition of f^{-1}. Since f is an isometry, $\|P' - Q'\| = \|f(P') - f(Q')\|$. In other words, $\left\|f^{-1}(P) - f^{-1}(Q)\right\| = \|P - Q\|$. This is exactly what we need to see that f^{-1} preserves distance. Since f^{-1} is also 1-1 and onto, it is an isometry. □

Let $\mathrm{Isom}(\mathbb{R}^2)$ be the set of all isometries of the plane $\mathbb{R}^2$. The lemma above shows that $\mathrm{Isom}(\mathbb{R}^2)$ is closed under the binary operation of function composition and under taking inverses. Associativity of composition holds automatically for functions from any set to itself, i.e., for every set A, functions $f, g, h : A \to A$ and every $a \in A$:

$$\begin{aligned} f \circ (g \circ h))(a) &= f(g(h(a))) \\ &= (f \circ g)(h(a)) \\ &= (f \circ g) \circ h(a) \end{aligned}$$

so that $f \circ (g \circ h)) = (f \circ g) \circ h$. Combined with the fact that the identity function is clearly an isometry, the lemma shows that $\mathrm{Isom}(\mathbb{R}^2)$ satisfies the axioms of a group.

An important class of isometries consists of the translations. To establish some notations that will be useful here and later in this chapter, fix a coordinate system on $\mathbb{R}^2$ with 0 representing the origin and associate to a point Q in $\mathbb{R}^2$ the displacement vector given by the directed line segment $\overrightarrow{0Q}$. In terms of coordinates, the point $Q = (a, b)$ is identified in this way with the vector $\overrightarrow{0Q} = \langle a, b \rangle$.

Translations

Definition 10.3. Let Q be a fixed point in $\mathbb{R}^2$ and $v = \langle a, b \rangle$ the vector identified with Q. The function $f(P) = P + v$ is the *translation* by the vector v.

It is trivial to see that f is an isometry. Its inverse is translation by $-v$. We will denote the translation by a vector v by τ_v. Clearly the inverse of τ_v is τ_{-v}. In terms of coordinates, if P is the point (x, y) and v is the vector $\langle a, b \rangle$ then $\tau_v(P) = (x + a, y + b)$.

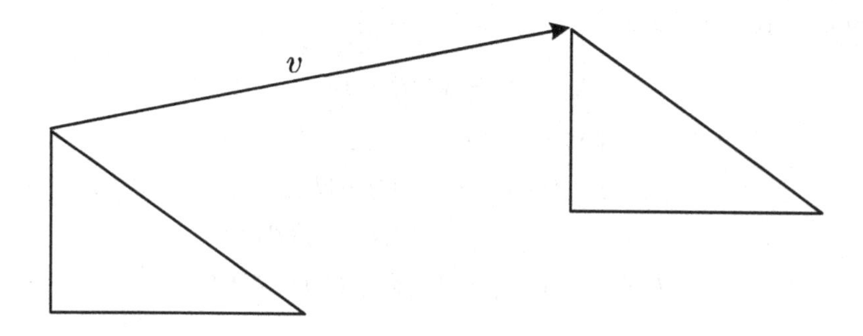

Lemma 10.4. *Let g be an isometry of $\mathbb{R}^2$ with $g(0) = v$. Then $\tau_{-v} \circ g := \tilde{g}$ is an isometry fixing the origin. In particular every isometry is the composition of an isometry fixing the origin with a translation.*

Proof. $g = (\tau_v \circ \tau_{-v}) \circ g = \tau_v \circ (\tau_{-v} \circ g) = \tau_v \circ \tilde{g}$. □

As the next proposition shows, the isometries fixing the origin are in fact linear transformations of $\mathbb{R}^2$, thus enabling the introduction of tools from linear algebra to the analysis of isometry groups. The *General Linear Group*, $\mathrm{GL}_n(\mathbb{R})$, and its subgroup the *Special Linear Group*, $\mathrm{SL}_n(\mathbb{R})$, play essential roles in what follows, so we remind the reader of their definitions.

Recall from Chap. 8 the *General Linear Group*, $\mathrm{GL}_n(\mathbb{R})$, whose elements are the nonsingular, i.e., invertible, linear transformations from $\mathbb{R}^n$ to itself with the group operation being composition. Equivalently, once a basis for $\mathbb{R}^n$ is specified, $\mathrm{GL}_n(\mathbb{R})$ is identified with the set of all nonsingular $n \times n$ matrices with entries in $\mathbb{R}$ with matrix multiplication as the group operation. Its subgroup, the *Special Linear Group*, $\mathrm{SL}_n(\mathbb{R})$, is the subgroup of $\mathrm{GL}_n(\mathbb{R})$ consisting of those matrices of determinant equal to 1. These groups play a special role in our discussion of isometries as the next proposition will show.

Lemma 10.5. *If f is an isometry of $\mathbb{R}^2$ and P, Q, R are collinear, then $f(P)$, $f(Q)$, and $f(R)$ are collinear.*

Proof. Let P, Q, R be collinear, labeled in collinear order, so that $\|R - Q\| + \|Q - P\| = \|R - P\|$. Since f is an isometry

$$\begin{aligned}\|f(R)-f(P)\| &= \|R-P\| \\ &= \|R-Q\| + \|Q-P\| \\ &= \|f(R)-f(Q)\| + \|f(Q)-f(P)\| .\end{aligned}$$

But $\|f(R)-f(P)\| = \|f(R)-f(Q)\| + \|f(Q)-f(P)\|$ implies that $f(P)$, $f(Q)$, and $f(R)$ are collinear, otherwise they are the vertices of a triangle for which $\|f(R)-f(P)\| < \|f(R)-f(Q)\| + \|f(Q)-f(P)\|$. □

Proposition 10.6. *Fix a coordinate system on* $\mathbb{R}^2$ *with origin denoted* 0. *The subset of* $\mathrm{Isom}(\mathbb{R}^2)$ *consisting of all isometries fixing* 0 *is a subgroup isomorphic to a subgroup of* $\mathrm{GL}_n(\mathbb{R})$, *i.e., the origin-preserving isometries are nonsingular linear transformations of the vector space* $\mathbb{R}^2$.

Proof. Let P, Q be points in $\mathbb{R}^2$ and $f \in \mathrm{Isom}(\mathbb{R}^2)$. The directed line through P, Q is parametrized as $\{(1-t)P + tQ : t \in \mathbb{R}\}$. For each $a \in \mathbb{R}$ set $X_a = (1-a)P + aQ$. By the previous lemma, $f(X_a)$ lies on the line joining $f(P)$ and $f(Q)$. Using the distance preserving property of isometries we have

$$\begin{aligned}\|X_a - P\| &= |a|||Q-P|| \\ \|X_a - Q\| &= |1-a|||Q-P|| \\ \|f(X_a) - f(P)\| &= |a|||(Q)-(P)|| \\ &= |a|||f(Q)-f(P)|| \\ \|f(X_a) - f(Q)\| &= |1-a|||f(Q)-f(P)||.\end{aligned}$$

If $0 \le a \le 1$, then X_a lies on the directed line segment joining P and Q at a distance a from P, and the same must be true for $f(X_a)$ with respect to the directed line segment joining $f(P)$ and $f(Q)$. Thus $f(X_a) = (1-a)f(P) + af(Q)$. The cases $a > 1$ and $a < 0$ are symmetric and handled similarly.

Taking $P = 0$, we have $X_a = aQ$, and by assumption $f(P) = 0$, so the multiplicative property $f(aQ) = af(Q)$ for every $a \in \mathbb{R}$ and every $Q \in \mathbb{R}^2$ holds. Taking arbitrary P and Q with $a = \frac{1}{2}$ we obtain

$$f\left(\frac{1}{2}(P+Q)\right) = \frac{1}{2}f(P) + \frac{1}{2}f(Q).$$

Applying the multiplicative property,

$$\begin{aligned}f(P+Q) &= 2\left[\frac{1}{2}f(P) + \frac{1}{2}f(Q)\right] \\ &= f(P) + f(Q).\end{aligned}$$

□

Corollary 10.7. *Every isometry of* $\mathbb{R}^2$ *preserves the angles between pairs of vectors.*

Proof. The assertion is clearly true for translations and by Lemma 10.4 we may restrict our attention to origin-preserving isometries, hence to isometries f which are linear transformations. Let v and w be vectors in $\mathbb{R}^2$ and $f \in \mathrm{Isom}(\mathbb{R}^2)$. The isometric property implies that $||f(v+w)|| = ||v+w||$. Recall that

$$\begin{aligned}||v+w||^2 &= (v+w)\cdot(v+w)\\ &= v\cdot v + v\cdot w + w\cdot v + w\cdot w\\ &= ||v||^2 + ||w||^2 + 2v\cdot w.\end{aligned}$$

Similarly, the additivity of f yields

$$\begin{aligned}||f(v+w)||^2 &= ||f(v)+f(w)||^2\\ &= ||f(v)||^2 + ||f(w)||^2 + 2f(v)\cdot f(w)\\ &= ||v||^2 + ||w||^2 + 2f(v)\cdot f(w),\end{aligned}$$

so that $f(v)\cdot f(w) = v\cdot w$. If $0 \le \theta \le \pi$ is the angle between v and w and $0 \le \varphi \le \pi$ the angle between $f(v)$ and $f(w)$, we have

$$\begin{aligned}f(v)\cdot f(w) &= ||f(v)||f||w||\cos\varphi = ||v||||w||\cos\varphi\\ &= v\cdot w = ||v||||w||\cos\theta\end{aligned}$$

from which we obtain $\cos\varphi = \cos\theta$ and therefore that $\varphi = \theta$. □

10.1.1 Origin-Preserving Isometries

From Proposition 10.6 we know that once a basis is chosen for $\mathbb{R}^2$ every origin-preserving isometry has a unique representation in $GL_n(\mathbb{R})$. The special features of matrices which represent isometries will be determined in this section.

Let $e_1 = \begin{pmatrix}1\\0\end{pmatrix}$ and $e_2 = \begin{pmatrix}0\\1\end{pmatrix}$ be the standard basis for $\mathbb{R}^2$ and $f \in \mathrm{Isom}(\mathbb{R}^2)$ with $f(0) = 0$. Write f in matrix form with respect to the standard basis

$$f = \begin{pmatrix}a & c\\ b & d\end{pmatrix}$$

so that $f(e_1) = \begin{pmatrix}a\\b\end{pmatrix}$, $f(e_2) = \begin{pmatrix}c\\d\end{pmatrix}$. Because e_1 and e_2 are orthogonal unit vectors, Corollary 10.7 implies the same for $\begin{pmatrix}a\\b\end{pmatrix}$ and $\begin{pmatrix}c\\d\end{pmatrix}$. In other words, the points (a,b) and (c,d) lie on the unit circle in $\mathbb{R}^2$, so that there is $0 \le \theta \le 2\pi$ with

$$\begin{aligned}(a,b) &= (\cos\theta, \sin\theta)\\ (c,d) &= \left(\cos\left(\theta \pm \frac{\pi}{2}\right), \sin\left(\theta \pm \frac{\pi}{2}\right)\right).\end{aligned}$$

Therefore the matrix representation for f is either the matrix

$$R_\theta = \begin{pmatrix}\cos\theta & -\sin\theta\\ \sin\theta & \cos\theta\end{pmatrix}.$$

of determinant equal to 1, or the matrix

$$f_\theta = \begin{pmatrix} \cos\theta & \sin\theta \\ \sin\theta & -\cos\theta \end{pmatrix}$$

of determinant equal to -1. Note that $f_\theta^2 = id$.

Another property of the matrices R_θ and f_θ is of essential importance: They are the orthogonal 2×2 matrices, i.e., their rows constitute pairs of orthogonal unit vectors ($R_\theta R_\theta^T = f_\theta f_\theta^T = I$), hence form an orthonormal basis for $\mathbb{R}^2$ (and the same is true of their columns as $R_\theta^T R_\theta = f_\theta^T f_\theta = I$ as well). Since we know that the collection of origin-preserving isometries is closed under composition and inversion, the collection of such matrices forms an important subgroup of $\mathrm{GL}_2(\mathbb{R})$ denoted $\mathrm{O}_2(\mathbb{R})$. More generally:

Definition 10.8. The *Orthogonal Group* $\mathrm{O}_n(\mathbb{R})$ is the subgroup of $\mathrm{GL}_n(\mathbb{R})$ consisting of those matrices whose rows (columns) constitute an orthonormal basis for $\mathbb{R}^n$. The subgroup $\mathrm{O}_n(\mathbb{R}) \cap \mathrm{SL}_n(\mathbb{R})$ of orthogonal matrices having determinant equal to 1 is called the *Special Orthogonal Group*, and is denoted $\mathrm{SO}_n(\mathbb{R})$.

That $\mathrm{O}_n(\mathbb{R})$ is a subgroup of $\mathrm{GL}_n(\mathbb{R})$ and $\mathrm{SO}_n(\mathbb{R})$ is a subgroup of $\mathrm{O}_n(\mathbb{R})$ are left as exercises. The isometries fixing the origin then fall into two classes:

Rotations

If θ is an angle, then the *rotation* (counterclockwise) by an angle θ about the origin is given in coordinates by the matrix

$$R_\theta =: \begin{pmatrix} \cos\theta & -\sin\theta \\ \sin\theta & \cos\theta \end{pmatrix}.$$

Observe that

$$R_\theta(x,y) = \left(\begin{pmatrix} \cos\theta & -\sin\theta \\ \sin\theta & \cos\theta \end{pmatrix}\begin{pmatrix} x \\ y \end{pmatrix}\right)^T = (x\cos\theta - y\sin\theta, x\sin\theta + y\cos\theta).$$

We can use this to describe a rotation about any point. If r' is the rotation by θ about a point $P \in \mathbb{R}^2$, and if t is the translation by P, then $r' = t \circ r \circ t^{-1}$. As a consequence, this shows that every rotation is an isometry. Note that r^{-1} is rotation by $-\theta$ about the origin. If $\theta = 2\pi/n$ for some integer n, then $r^n = \mathrm{id}$.

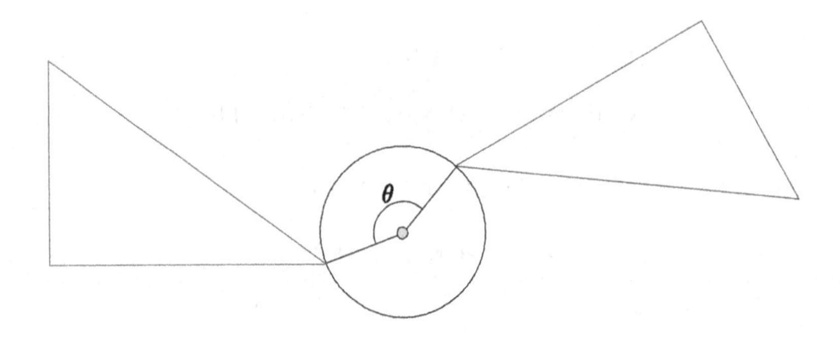

Reflections

Let ℓ be the line in $\mathbb{R}^2$ passing through the origin at angle $\frac{\theta}{2}$ with the vector e_1, i.e., ℓ is parallel to the vector $\langle \cos\frac{\theta}{2}, \sin\frac{\theta}{2}\rangle$. The map f that reflects points across ℓ is an isometry preserving the origin, so is a linear transformation. To find its matrix, we calculate the unit vectors $f(e_1) = u_1$ and $f(e_2) = u_2$ (f is an isometry). Since the angle between ℓ and u_1 is also equal to $\frac{\theta}{2}$ we have $u_1 = \langle \cos\theta, \sin\theta\rangle$. Write $u_2 = \langle a, b\rangle$. Since f reflects the point $(0, 1)$ to its mirror image about ℓ, this line is the perpendicular bisector of the line segment ℓ' joining $(0, 1)$ and (a, b). In particular the slope of ℓ' is equal to $-\cot\frac{\theta}{2}$ so that it has equation $y - 1 = -\cot\frac{\theta}{2}x$. Precisely two points on ℓ' are at distance 1 from the origin:

$$\begin{aligned}
1 &= x^2 + (1 - \cot\frac{\theta}{2}x)^2 \\
&= x^2 + 1 - 2\cot\frac{\theta}{2}x + \cot^2\frac{\theta}{2}x^2 \\
&= x^2\left(1 + \cot^2\frac{\theta}{2}\right) - 2\cot\frac{\theta}{2}x + 1,
\end{aligned}$$

which simplifies to

$$0 = x\left(x\csc^2\frac{\theta}{2} - 2\cot\frac{\theta}{2}\right).$$

The solution $x = 0$ yields the point $(0, 1)$, while the other

$$\begin{aligned}
x &= \frac{2\cot\frac{\theta}{2}}{\csc^2\frac{\theta}{2}} \\
&= \frac{2\sin^2\frac{\theta}{2}\cos\frac{\theta}{2}}{\sin\frac{\theta}{2}} \\
&= 2\sin\frac{\theta}{2}\cos\frac{\theta}{2} \\
&= \sin\theta
\end{aligned}$$

yields the point $(\sin\theta, -\cos\theta)$. Thus the matrix of f with respect to the standard basis is exactly the matrix f_θ above, in particular reflection across a line is an isometry satisfying $f^2 = \text{id}$.

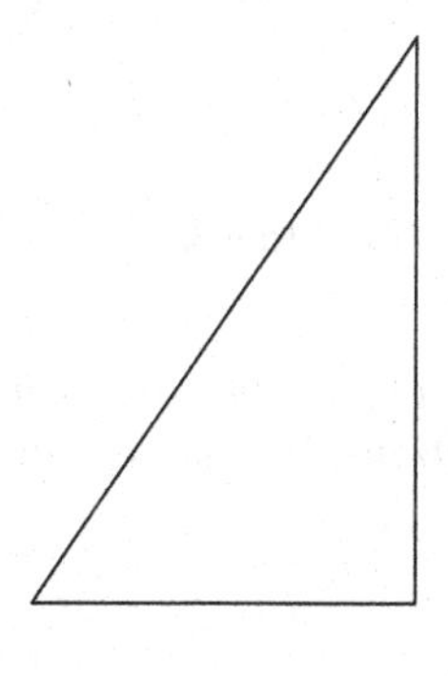

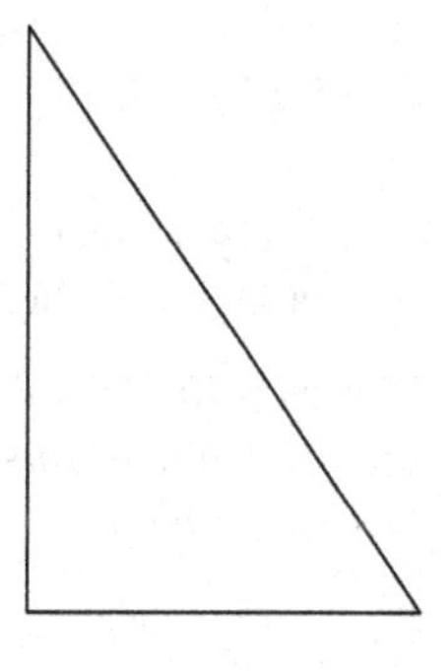

Alternatively, one can verify that if ℓ is the line through the origin parallel to a vector w, then the *reflection* across ℓ of the vector v is given by the formula

$$f(v) = 2\left(\frac{v \cdot w}{w \cdot w}\right) w - v.$$

This formula comes from the formula for projection of one vector onto another that one sees in multivariable calculus.

Proposition 10.9. *Every isometry is determined by the images of any 3 non-collinear points in* $\mathbb{R}^2$.

Proof. Let P, Q, R be non-collinear points and $f \in \mathrm{Isom}(\mathbb{R}^2)$. Choose coordinates so that the origin $O = P$, set $V = f(O)$, and $v = \overrightarrow{OV}$. Then $t_{-v} \circ f$ fixes O and is therefore a linear transformation. By non-collinearity, the displacement vectors $\overrightarrow{OQ}$ and $\overrightarrow{OR}$ are linearly independent so that $\tau_{-v} \circ f$ is determined by its values on these two vectors, i.e., by $\tau_{-v} \circ f(R)$ and $\tau_{-v} \circ f(Q)$. Since $f = \tau_v \circ \tau_{-v} \circ f$, this isometry is determined by $f(O), f(Q)$, and $f(R)$. □

10.1.2 Compositions of Isometries

10.1.2.1 Glide Reflections

Since $\mathrm{Isom}(\mathbb{R}^2)$ is a group under function composition, new isometries may be produced this way. It is clear that the composition of two rotations about the same point is again a rotation (about the same point). In the exercises below, the composition of a rotation and a translation is seen to be a rotation, and the composition of a rotation and a reflection is shown to result in either a rotation or a translation. The composition of a reflection and a translation may be a reflection, although it may also be a new type of isometry. We will call such a composition a *glide reflection* if it is not a reflection.

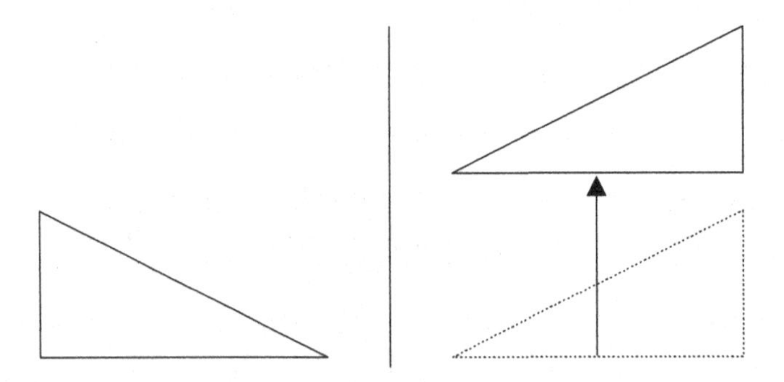

Since, by Lemma 10.4, every isometry is the composition of a translation with an isometry that preserves the origin, we have proved the following theorem:

Theorem 10.10. *Every isometry of* $\mathbb{R}^2$ *is a composition of a translation with either a rotation or a reflection. Therefore translations, rotations, reflections, and glide reflections account for all types of isometries of the plane.*

10.2 Structure of the Group Isom($\mathbb{R}^2$)

In this section the structure theory introduced in Chap. 9 is applied to the group of isometries of $\mathbb{R}^2$. The need for those results is illustrated by the following observation. Recall the matrices R_θ and f_θ representing, respectively, rotation about the origin through the angle θ and reflection about the line through the origin parallel to the vector $\langle \cos\frac{\theta}{2}, \sin\frac{\theta}{2}\rangle$. Calculating the product $R_\theta f_\theta$ results in the matrix $f_{2\theta}$ while $f_\theta R_\theta = f_\pi$. The group Isom($\mathbb{R}^2$) is therefore nonabelian. Nevertheless it is possible to understand its structure and application to symmetries of planar regions, by studying the subsets consisting of the translations and of the origin-preserving isometries. It is clear that the composition of two translations and the inverse of a translation are again translations. Similarly the property of fixing the origin is preserved under composition and taking inverses. This observation is recorded as

Proposition 10.11. *The collections T of translations and P of origin-preserving isometries are subgroups of* Isom($\mathbb{R}^2$).

The subgroup T has some special properties. It is an abelian group, i.e., composition of translations is commutative, which follows because $\tau_v \circ \tau_w = \tau_{v+w}$, and vector addition is commutative. The origin-preserving isometries constitute a nonabelian subgroup (Exercise 4). More importantly for the structure of Isom($\mathbb{R}^2$) we have:

Lemma 10.12. *Let T be a translation and f an arbitrary isometry. Then $f^{-1} \circ T \circ f$ is again a translation. In particular T is a normal subgroup of* Isom($\mathbb{R}^2$).

Proof. Choose a coordinate system with origin O, set $V = f(O)$, $v = \overrightarrow{OV}$ and $g = \tau_{-v} \circ f$. Then $f = \tau_v \circ g$ and

$$\begin{aligned} f^{-1}Tf &= (\tau_v \circ g)^{-1} \circ T \circ \tau_v \circ g \\ &= g^{-1} \circ \tau_{-v} \circ T \circ \tau_v \circ g \\ &= g^{-1} \circ T \circ g \end{aligned}$$

so we may assume that f is origin-preserving, hence a linear transformation with matrix A. In coordinates then, with $T = t_w$, we compute

$$\begin{aligned} f^{-1}Tf(x,y) &= A^{-1}(A(x,y)^T + w) \\ &= (x,y) + A^{-1}w, \end{aligned}$$

i.e., $f^{-1}Tf = t_{A^{-1}w}$ the translation by $A^{-1}w$. □

In contrast, the subgroup of origin-preserving isometries is not normal. For a simple example, let f be the counterclockwise rotation about the origin through $\frac{\pi}{2}$ radians and $v = \langle 1, 0\rangle$. Then

$$\begin{aligned} \tau_{-v} \circ f \circ \tau_v(x,y) &= \begin{pmatrix} -1 \\ 0 \end{pmatrix} + \begin{pmatrix} 0 & -1 \\ 1 & 0 \end{pmatrix}\begin{pmatrix} x+1 \\ y \end{pmatrix} \\ &= \begin{pmatrix} -y-1 \\ x+1 \end{pmatrix}, \end{aligned}$$

which certainly does not fix the origin.

10.2.1 Semidirect Products

With respect to the subgroups T and P of Isom($\mathbb{R}^2$), we found that T is normal, Isom($\mathbb{R}^2$) $=TP$ and $T \cap P = \{e\}$. However, because P is not a normal subgroup, i.e., Condition 2 of Theorem 9.19 is not satisfied, Isom($\mathbb{R}^2$) does not decompose as the direct product of the subgroups T and P. Nevertheless, as will be shown later in this section, there is a closely related decomposition of Isom($\mathbb{R}^2$). To apply these results to the structure of Isom($\mathbb{R}^2$), recall that every isometry g has a unique representation as $g = \tau \circ f$ with $\tau \in T$ and $f \in P$, but it is not the case that Isom($\mathbb{R}^2$) $\cong T \times P$ since, as we have seen, the elements of T and P do not in general commute. The failure of commutativity of $\tau \in T$ and $f \in P$ is equivalent to $f^{-1} \circ \tau \circ f \neq \tau$ but normality of T implies that $f^{-1} \circ \tau \circ f \in T$.

Recall from Example 9.5 that for $f \in$ Isom($\mathbb{R}^2$) the function c_f : Isom($\mathbb{R}^2$) $\to$ Isom($\mathbb{R}^2$), given by $c_f(g) = f^{-1} \circ g \circ f$, is an automorphism of Isom($\mathbb{R}^2$) and is of particular importance. In fact this type of mapping is important in the general theory of groups, so we return to our general development of group theory and later apply the results to the structure of Isom($\mathbb{R}^2$).

Definition 10.13. Let G be a group and $f \in G$. The function $c_f : G \to G$ given by $c_f(g) = f^{-1} \circ g \circ f$ is called *conjugation* by f.

The following proposition formalizes the content of Example 9.5.

Proposition 10.14. *Let G be a group. For every $f \in G$ the conjugation map $c_f(g) = f^{-1} \circ g \circ f$ is an isomorphism of groups, hence an automorphism of the group G.*

The condition of normality of a subgroup is conveniently expressed in terms of conjugation automorphisms: If N is a normal subgroup of G, then $c_g(n) = gng^{-1} \in N$ for every $g \in G$ and $n \in N$. More precisely, we have the following corollary. The proof, which is left to the exercises, uses the same reasoning as in Example 9.5.

Corollary 10.15. *The subgroup N of the group G is normal if and only if c_f restricts to an automorphism of N for every f in G.*

We now come to the promised decomposition of Isom($\mathbb{R}^2$). It relies on a more general version of the method used in Chap. 9 for putting a group structure on the Cartesian product of two groups.

Definition 10.16. Let G be a group with subgroups H and N satisfying the following conditions:

1. $G = HN$,
2. N is a normal subgroup of G, and
3. $H \cap N = \{e_G\}$.

Then G is said to be the semidirect product of N by H.

Remark 10.17. With this terminology, Isom($\mathbb{R}^2$) is the semidirect product of T by P.

The condition $H \cap N = \{e_G\}$ implies that the representation of $g \in G$ as $g = hn$ with $h \in H$, $n \in N$ is unique. Therefore the semidirect product can be interpreted as a group structure on the set $H \times N$. Indeed if $h_1 n_1 = h_2 n_2$, then $h_2^{-1} h_1 = n_2^{-1} n_1 \in H \cap N = \{e_G\}$ so that $h_1 = h_2$ and $n_1 = n_2$. The product in G can then be realized as

$$\begin{aligned} h_1 n_1 h_2 n_2 &= h_1 h_2 h_2^{-1} n_1 h_2 n_2 \\ &= (h_1 h_2)(h_2^{-1} n_1 h_2) n_2, \end{aligned}$$

which has the required form by the normality of N. If, in addition, H is a normal subgroup, then the elements of H and N commute and the semidirect product is in fact the direct product of H and N. For $h \in H$, and $n \in N$ note that with both H and N normal

$$\begin{aligned} hnh^{-1}n^{-1} &= (hnh^{-1})n^{-1} \in N \\ &= h(nh^{-1}n^{-1}) \in H, \end{aligned}$$

so that $hnh^{-1}n^{-1} = e_G$, i.e., $hn = nh$.

Remark 10.18. The natural group operation on the subgroup $T \cong \mathbb{R}^2$ of Isom($\mathbb{R}^2$) is addition while that on $P \cong O_2(\mathbb{R})$ is by multiplication. Taking advantage of the unique representation of elements of Isom($\mathbb{R}^2$) as $f_A \circ t_a$ with $A \in O_2(\mathbb{R})$ and $a \in \mathbb{R}^2$, the group structure on Isom($\mathbb{R}^2$) can be viewed as the one on $O_2(\mathbb{R}) \times \mathbb{R}^2$ given by

$$\begin{aligned} (A, a) \cdot (B, b) &= (AB, B^{-1}a + b) \\ (A, a)^{-1} &= (A^{-1}, -Aa). \end{aligned}$$

Proposition 10.19. *Let G be the semidirect product of N by H. Then $G/N \cong H$. In particular* Isom($\mathbb{R}^2$)$/T \cong P$.

Proof. Since every element $g \in G$ is uniquely represented as $g = hn$ with $h \in H$, $n \in N$, we can define a function $\pi : G \to H$ by $\pi(g) = h$. Clearly π is onto. We show that π is a group homomorphism with kernel equal to N. To that end, let $g_1 = h_1n_1$ and $g_2 = h_2n_2$ be elements of G with $h_1, h_2 \in H$, and $n_1, n_2 \in N$. Then

$$\begin{aligned} g_1g_2 &= h_1n_1h_2n_2 \\ &= h_1h_2h_2^{-1}n_1h_2n_2 \\ &= (h_1h_2)(h_2^{-1}n_1h_2n_2). \end{aligned}$$

The second factor is in N by normality so this is the unique expression of g_1g_2 as the product of an element of H with one an element of G. Thus

$$\pi(g_1g_2) = h_1h_2 = \pi(g_1)\pi(g_2)$$

and π is a homomorphism. The kernel of π is $\{e_Gn : n \in N\} = N$. Finally Theorem 9.25 implies that $G/N \cong H$. □

Exercises

1. Prove that the composition of a rotation and a translation is again a rotation.
2. Using $f(v) = 2\left(\frac{v \cdot w}{w \cdot w}\right) w - v$ for the *reflection* across the line ℓ of the vector v, recover the matrix representation of the reflection across the line through the origin parallel to the vector $\left(\cos\frac{\theta}{2}, \sin\frac{\theta}{2}\right)$ that was carried out above using trigonometry and analytic geometry.
3. Let f_1 and f_2 be reflections about the lines ℓ_1 and ℓ_2, respectively. Suppose that ℓ_1 and ℓ_2 are parallel and that b is the vector perpendicular to these lines such that translation by b sends ℓ_1 to ℓ_2. Show that f_2f_1 is translation by $2b$.

4. Determine the two compositions $f_{-\frac{\pi}{4}} \circ f_{\frac{\pi}{4}}$ and $f_{\frac{\pi}{4}} \circ f_{\frac{-\pi}{4}}$ and interpret the resulting isometries geometrically.
5. Let r be a rotation about a point P and let f be a reflection. Prove that fr is a rotation about P if $f(P) = P$, but that fr is a translation if $f(P) \neq P$.
6. Let f be a reflection about a line ℓ and let τ be a translation by a vector b. If b is parallel to ℓ, show that τf is a reflection about the line $\ell + b/2$.
7. Let f be a reflection about a line ℓ and let τ be a translation by a vector b. If b is not parallel to ℓ, show that τf fixes no point, so is not a reflection. Moreover, if $b = b_1 + b_2$ with b_1 parallel to ℓ and b_2 perpendicular to ℓ, show that τf is the composition $\tau' f'$ of the translation τ' by b_2 and f' is the reflection about the line $\ell + b_1/2$.
   (By this problem, it follows that any glide reflection is the composition of a reflection followed by a translation by a vector perpendicular to the reflection line.)
8. Demonstrate that the group of Example 9.14 is the semidirect product of a cyclic subgroup of order three with a cyclic subgroup of order 2.
9. Let $p > 2$ be a prime number. Prove that every group of order $2p$ is either the direct product or the semidirect product of a of order p by a group of order 2.
10. Let G be the semidirect product of a normal subgroup N and a subgroup H. Following the notation above, for each $h \in H$, we have the map $c_h : N \to N$ sending n to hnh^{-1}, which is an automorphism of N. Show that the function $H \to \mathrm{Aut}(N)$ given by $h \mapsto c_h$ is a group homomorphism.
11. Prove that $\mathrm{O}_n(\mathbb{R})$ is a subgroup of $\mathrm{GL}_n(\mathbb{R})$.
12. Prove that $\mathrm{SO}_n(\mathbb{R})$ is a normal subgroup of $\mathrm{O}_n(\mathbb{R})$.
13. Since $\mathrm{GL}_1(\mathbb{R}) = \mathbb{R}^*$ and $\mathrm{O}_1(\mathbb{R}) = \{1, -1\}$ under multiplication, $\mathrm{O}_1(\mathbb{R})$ is a normal subgroup of $\mathrm{GL}_1(\mathbb{R})$. Is $\mathrm{O}_2(\mathbb{R})$ a normal subgroup of $\mathrm{GL}_2(\mathbb{R})$? If so, prove it. If not show via example why not.
14. Prove that $\mathrm{O}_n(\mathbb{R})/\,\mathrm{SO}_n(\mathbb{R}) \cong \mathbb{Z}_2$.
15. Apply the methods in Example 9.5 to prove Corollary 10.15.

10.3 Symmetry Groups

Let X be a subset of the plane. We associate a group to X, called the *symmetry group* of X. This notion makes perfect sense for subsets of $\mathbb{R}^n$ for any n; but, we will restrict our attention to plane figures.

Definition 10.20. If X is a subset of the plane, then the symmetry group of X is the set of all $f \in \mathrm{Isom}(\mathbb{R}^2)$ for which $f(X) = X$. This group is denoted by $\mathrm{Sym}(X)$.

To better understand the definition, we look at it more carefully. If f is an isometry and X is a subset of the plane, then $f(X) = \{f(P) : P \in X\}$. Therefore, $f \in \mathrm{Sym}(X)$ if for every point $P \in X$, we have $f(P) \in X$ and, for every $Q \in X$ there is a $P \in X$ with $f(P) = Q$. We do not require that f restricts to the identity on X, i.e., that $f(P) = P$ for $P \in X$. For example, if $X = \{(0,1), (1,0), (-1,0), (0,-1)\}$,

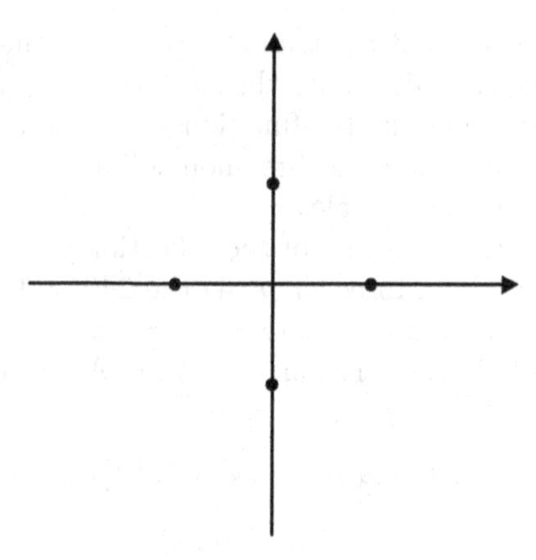

then a rotation of 90° about the origin sends this set of four points to itself. Thus, this rotation is a symmetry of X. To see, in general, that $\mathrm{Sym}(X)$ is a subgroup of $\mathrm{Isom}(\mathbb{R}^2)$, we only need to show that if $f, g \in \mathrm{Sym}(X)$, then $f \circ g^{-1} \in \mathrm{Sym}(X)$. However, if $f, g \in \mathrm{Sym}(X)$, then $f(X) = g(X) = X$. Therefore, $g^{-1}(X) = X$, and so $f(g^{-1}(X)) = f(X) = X$.

10.3.1 Examples of Symmetry Groups

The examples we give in this subsection will help us to understand the definition of symmetry group and to introduce as symmetry groups two important classes of groups, the cyclic and dihedral groups.

Example 10.21. We calculate the symmetry group of an equilateral triangle T. For convenience, we view the origin as the center of the triangle.

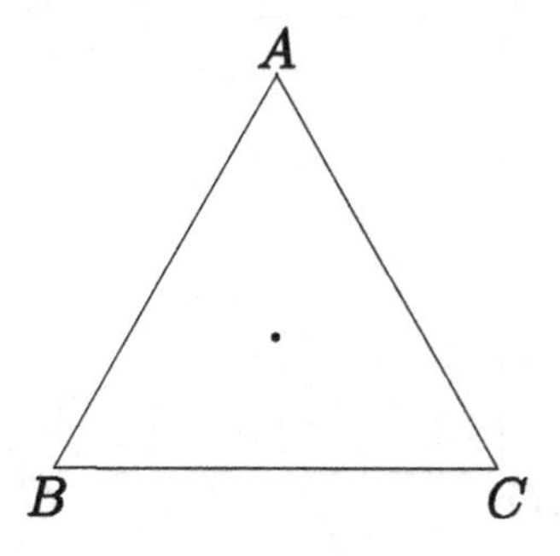

The identity function is always a symmetry of any figure. Also, the rotations about the origin by an angle of 120° or 240° are elements of $\mathrm{Sym}(T)$. Furthermore, the three reflections across the three dotted lines of the following picture are also isometries.

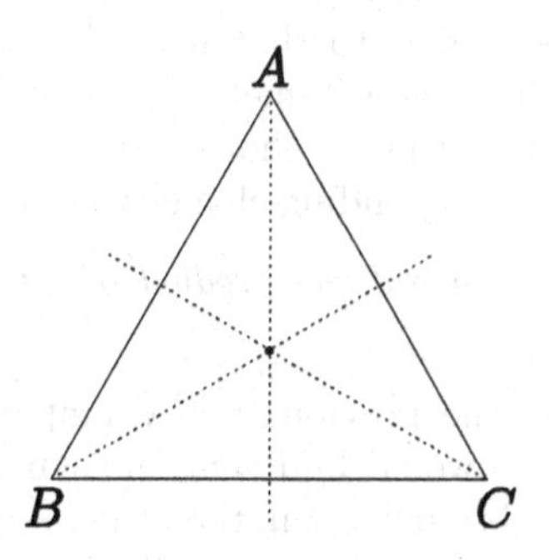

We have so far found six isometries of the triangle. We claim that $\mathrm{Sym}(T)$ consists precisely of these six isometries. To see this, first note that any isometry must send the set $\{A, B, C\}$ of vertices to itself. There are only six 1-1 functions that send $\{A, B, C\}$ to itself, and any isometry is determined by what it does to three non-collinear points by Proposition 10.9, so we have found all symmetries of the triangle.

If r is the rotation by 120° and f is any of the reflections, then an exercise will show that f, rf, r^2f are the three reflections. Moreover, r^2 is the 240° rotation. Therefore, $\mathrm{Sym}(T) = \{e, r, r^2, f, rf, r^2f\}$. The elements r and f satisfy $r^3 = e$ and $f^2 = e$. Another exercise will show that r and f are related via the relation $fr = r^2f$. Alternatively, this equation may be rewritten as $frf = r^{-1}$, since $f^{-1} = f$ and $r^2 = r^{-1}$.

Corollary 10.22. $\mathrm{Sym}(T)$ *is the semidirect product of the normal subgroup* $\{e, r, r^2\}$ *by the subgroup* $\{e, f\}$.

Remark 10.23. $\mathrm{Sym}(T)$ is isomorphic to the group in Example 9.14.

Example 10.24. Now let us determine the symmetry group of a square Q, centered at the origin for convenience. As with the triangle, we see that any isometry that preserves the square must permute the four vertices. There are 24 permutations of the vertices. However, not all come from isometries. First, the rotations of 0°, 90°, 180°, and 270° about the origin are symmetries of Q. Also, the reflections about the four dotted lines in the picture below are also symmetries. We now have eight symmetries of the square and we claim that this is all. To see this, we give a counting argument. There are four choices for where vertex A can be sent since it must go to one of the vertices. Once a choice has been made, there are just two choices for where B is sent since it must go to a vertex adjacent to the image of A; this is forced upon us since isometries preserve distance. After images for A and B have been chosen, the images of the other two vertices are then fixed; C is sent to the vertex across from the image of A, and

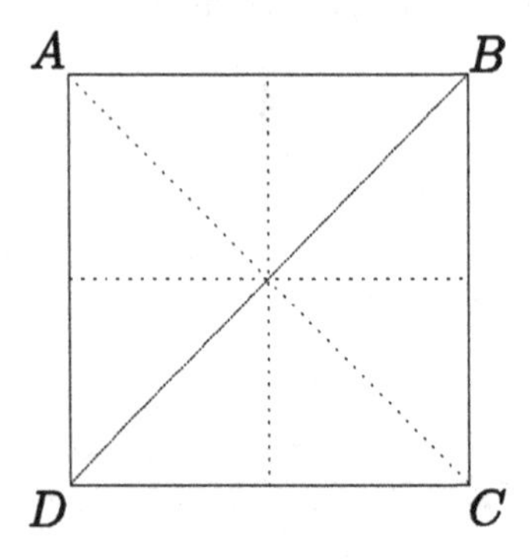

D is sent to the vertex across from the image of B. So, there is a total of $4 \cdot 2 = 8$ possible isometries. Since we have found eight, we have them all.

If r is the rotation by 90° and f is any reflection, then the four reflections are f, rf, r^2f, r^3f; this can be seen by an exercise similar to that needed in the previous example. The four rotations are $r, r^2, r^3, r^4 = e$. Thus, the symmetry group of Q is $\{e, r, r^2, r^3, f, rf, r^2f, r^3f\}$. We have $r^4 = e$ and $f^2 = e$, and an exercise shows that $fr = r^3f$, or $frf = r^{-1}$. This is the same relation that holds for the corresponding elements in the previous example.

Corollary 10.25. $\mathrm{Sym}(Q)$ *is the semidirect product of the normal subgroup* $\{e, r, r^2, r^3\}$ *by the subgroup* $\{e, f\}$.

Example 10.26. If we generalize the previous two examples by considering the symmetry group of a regular n-gon, then we would find that the symmetry group has $2n$ elements. If r is a rotation by $360°/n$ and if f is a reflection, then the rotations in the group are the powers

$r, r^2, \ldots, r^{n-1}, r^n = e$ of r, and the reflections are $f, rf, \ldots, r^{n-1}f$. Thus, half of the elements are rotations and half are reflections. The elements r and f satisfy the relations $r^n = e$, $f^2 = e$, and $frf = r^{-1}$.

Definition 10.27. The *Dihedral group* of order $2n$, denoted D_n, is the group of symmetries of the regular n–gon.

Because of the relation $frf = r^{-1}$, the elements of D_n can be listed as $e, r, r^2, \ldots, r^{n-1}, f, rf, r^2f, \ldots, r^{n-1}f$. The subgroup of all rotations in D_n is also of importance for us. This is the cyclic subgroup $\langle r \rangle = \{e, r, \ldots, r^{n-1}\}$ denoted by C_n.

Theorem 10.28. *For every $n \geq 3$, D_n is the semidirect product of C_n by $\{e, f\}$.*

Proof. It is clear that $D_n = C_n \langle f \rangle$ and $C_n \cap \langle f \rangle = \{e\}$. Normality of the subgroup C_n follows from Proposition 9.13. □

The group C_n also arises as a symmetry group. The following example shows how C_4 can arise as a symmetry group.

Example 10.29. The following figure has only rotations in its symmetry group; because it has rotations only of 0°, 90°, 180°, and 270°, its symmetry group is C_4.

By drawing similar but more complicated pictures, we can represent C_n as a symmetry group of some plane figure.

Example 10.30. The Zia symbol, which appears in the state flag of New Mexico, has symmetry group D_4; we see that besides rotations by 0°, 90°, 180°, and 270°, it has horizontal, vertical, and diagonal reflections in its symmetry group, and no other symmetry.

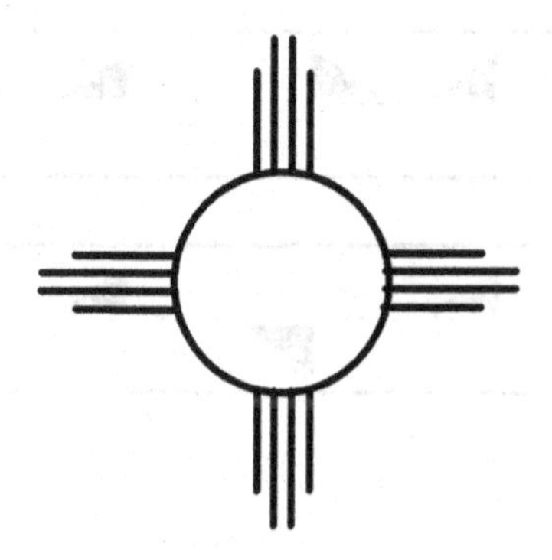

Example 10.31. Consider a circle of radius 1 centered at the origin $\mathbf{0}$. Any isometry of the circle must map the center to itself. Thus, any such isometry is linear, by Proposition 10.6. Conversely, if φ is any linear isometry, and if P is any point with $d(P, \mathbf{0}) = 1$, then $d(\varphi(P), \varphi(\mathbf{0})) = d(\varphi(P), \mathbf{0}) = 1$. Thus, φ sends the circle to itself. Therefore, the symmetry group of the circle is $O_2(\mathbb{R})$.

Exercises

1. Let G be the symmetry group of an equilateral triangle. If r is a 120° rotation and f is a reflection, show that f, rf, and $r^2 f$ are the three reflections of G.
2. Show that there are exactly six 1-1 functions from a set of 3 elements to itself.
3. Let G be the symmetry group of an equilateral triangle. If r is a 120° rotation and f is a reflection, so that $fr = r^2 f$. Use this to show that $frf = r^{-1}$.
4. Let G be the symmetry group of a square. If r is a 90° rotation and f a reflection, show that the four reflections of G are $f, rf, r^2 f$, and $r^3 f$.
5. Let G be the symmetry group of a square. If r is a 90° rotation and f a reflection, show that $fr = r^3 f$. Use this to show that $frf = r^{-1}$.

10.4 The Seven Frieze Groups

Frieze patterns are often found as decorative borders, for instance, on fabrics or along walls. A frieze pattern is a one-dimensional "tiling" consisting of a figure that is repeated at regular intervals in one direction. For example,

is a frieze pattern, as are:

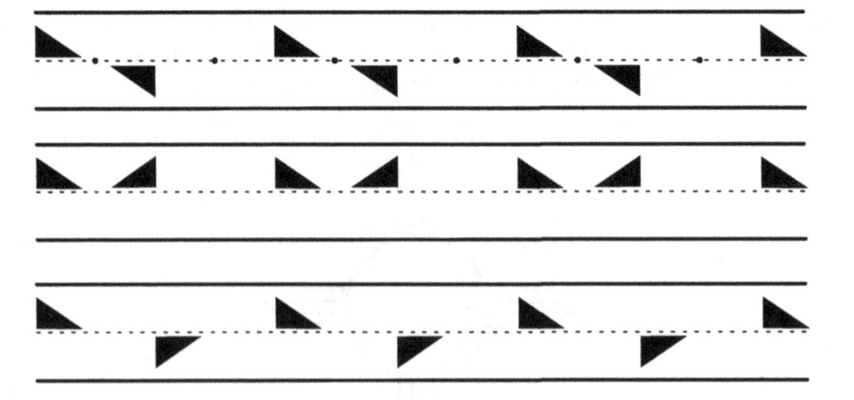

Since frieze patterns are subsets of the plane we can investigate their symmetry groups, which are called frieze groups. It turns out that there are only seven possible groups of isometries that can arise as symmetry groups of frieze patterns. Although human artistry can produce innumerable varieties of frieze patterns, from a mathematical point of view there are only seven different types.

For concreteness assume, as in the examples above, that our frieze patterns extend infinitely in the horizontal direction. If $\mathcal{F}$ is a frieze pattern, denote its symmetry group by $\mathrm{Sym}(\mathcal{F})$. By definition, $\mathrm{Sym}(\mathcal{F})$ contains a translation T of some minimal length L in the horizontal direction (e.g., the length between two consecutive triangles above the dotted line in the first two examples, respectively). Therefore $\mathrm{Sym}(\mathcal{F})$ contains as a subgroup $\{nT_L : n \in \mathbb{Z}\} \cong \mathbb{Z}$ and this group must equal $\mathrm{Sym}(\mathcal{F}) \cap T$.

Choose one tile and designate its center as the origin of a coordinate system on $\mathbb{R}^2$. We consider the possibilities for $\mathrm{Sym}(\mathcal{F}) \cap P$. Since $\mathcal{F}$ extends in the horizontal direction, the only reflections available are those about a vertical line through the origin, denoted f_v (as in the third example above) and about a horizontal line through the origin, denoted f_h. Similarly, the only nontrivial rotation available is a rotation r about the origin through 180° (as in the second example above). Recall that $f_v \circ f_h = f_h \circ f_v = r$. There is only one glide reflection available, namely $f_h \circ T_{\frac{L}{2}}$, which is a symmetry of the fourth example above.

Our objective us to show that there are only seven distinct subgroups of $\mathrm{Isom}(\mathbb{R}^2)$ containing T_L and some of f_v, f_h, r, and $f_h \circ T_{\frac{L}{2}}$. For simplicity set $g_h = f_h \circ T_{\frac{L}{2}}$, and e equal to the identity symmetry. Since a group is closed under its operation, the following multiplication table for these isometries is helpful in identifying the restrictions on a pattern $\mathcal{F}$ imposed by its symmetries. For instance, the equation $r \circ g_h = f_v \circ T_{\frac{L}{2}}$ says that a frieze symmetric with respect to both r and g_h must be very symmetric.

$\circ$	f_v	f_h	r	g_h
f_v	e	r	f_h	$r \circ T_{\frac{L}{2}}$
f_h	r	e	f_v	$T_{\frac{L}{2}}$
r	f_h	f_v	e	$r \circ g_h = f_v \circ f_h \circ f_h \circ T_{\frac{L}{2}} = f_v \circ T_{\frac{L}{2}}$
g_h	$r \circ T_{\frac{L}{2}}$	$T_{\frac{L}{2}}$	$f_v \circ T_{\frac{L}{2}}$	T_L

If G is to be the symmetry group of a frieze pattern, G must contain $\{nT_L : n \in \mathbb{Z}\}$, some subset of $\{f_v, f_h, r, g_h\}$, and must be closed under composition. From the table above, we see that G that there can only be the following seven possibilities, given with frieze patterns exhibiting precisely those symmetry groups. For elements $a_1, a_2, \ldots, a_n$ of a given group $\mathcal{G}$ we write $\langle a_1, a_2, \ldots, a_n \rangle$ for the smallest subgroup of $\mathcal{G}$ containing $\{a_1, a_2, \ldots, a_n\}$.

1. $G = \langle T_L \rangle$

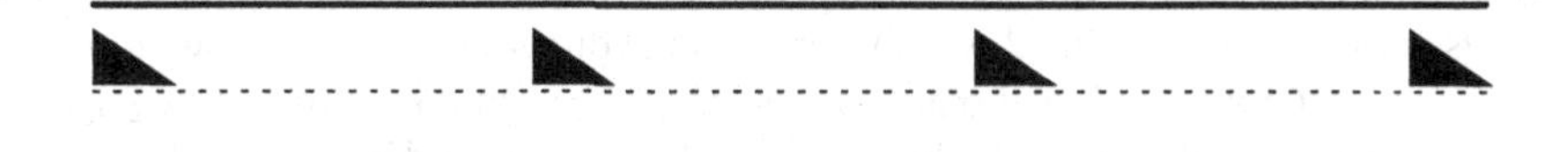

2. $G = \langle T_L, f_h \rangle$

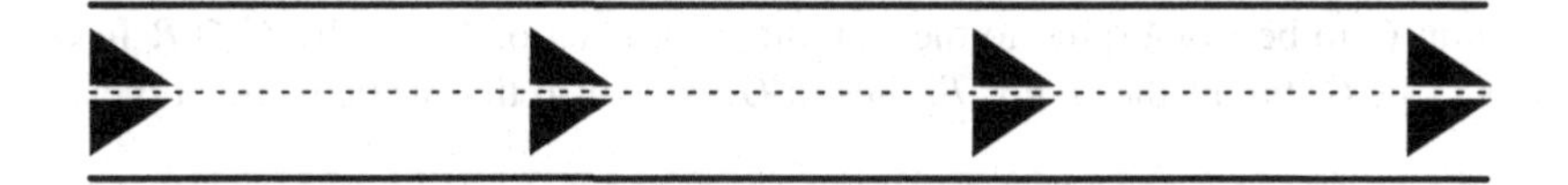

3. $G = \langle T_L, f_v \rangle$

4. $G = \langle T_L, r \rangle$

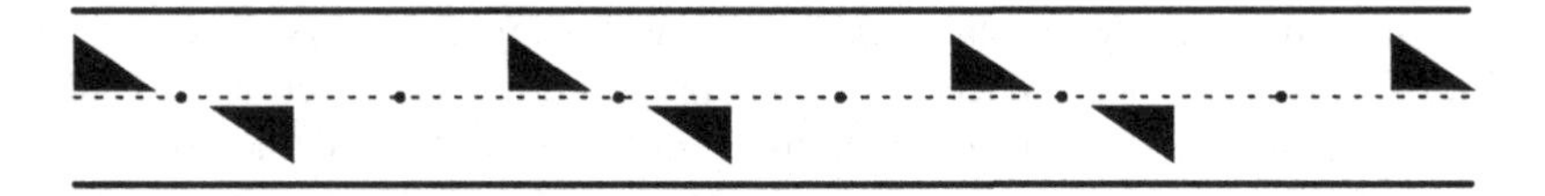

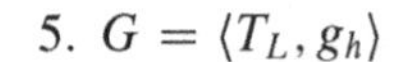

5. $G = \langle T_L, g_h \rangle$

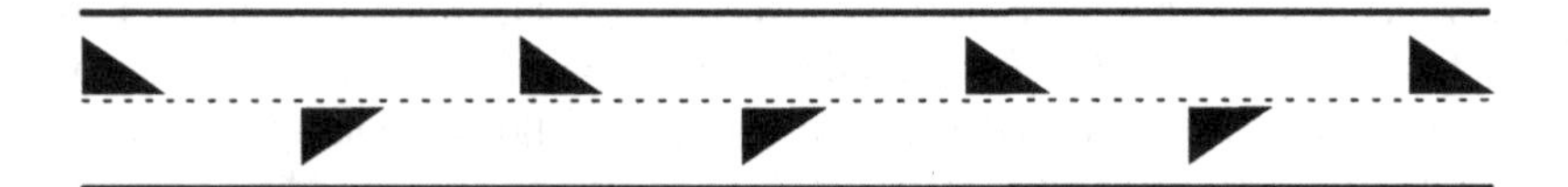

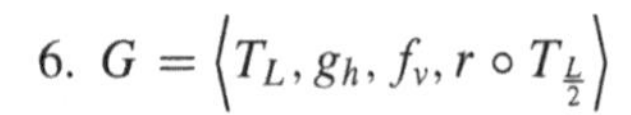

6. $G = \left\langle T_L, g_h, f_v, r \circ T_{\frac{L}{2}} \right\rangle$

7. $G = \langle T_L, f_h, f_v, r \rangle$

Of these groups only groups 1, 2, and 5 are abelian. Group 1 is isomorphic to the group $\mathbb{Z}$ of integers under addition. So is Group 5 because $g_h \circ g_h = T_L$. Thus $\langle T_L, g_h \rangle = \langle g_h \rangle$ an infinite cyclic group, hence isomorphic to $\mathbb{Z}$. Group 2 is isomorphic to the direct product of the subgroups $\langle T_L \rangle$ and $\langle f_h \rangle$ hence to $\mathbb{Z} \times \mathbb{Z}_2$. Groups 3, 4, and 6 are isomorphic to the semidirect of $\mathbb{Z}$ by $\mathbb{Z}_2$ specifically to the semidirect products of subgroup $\langle T_L \rangle$ by $\langle f_v \rangle$, of $\langle T_L \rangle$ by $\langle r \rangle$, and of $\langle g_h \rangle$ by $\langle f_v \rangle$, respectively. Finally, Group 7 is isomorphic to the direct product of the subgroup $\langle f_h \rangle \cong \mathbb{Z}_2$ and the subgroup $\langle T_L, f_v \rangle$, the latter of which we have noted is isomorphic to the semidirect product of $\mathbb{Z}$ by $\mathbb{Z}_2$.

Although symmetry groups of plane figures are subgroups of Isom($\mathbb{R}^2$) which, as we have seen, has the structure of the semidirect product of the subgroups T and P, one cannot expect an arbitrary symmetry group G to be isomorphic to the semidirect product of $G \cap T$ by $G \cap P$. In Group 6 for example, $r \circ T_{\frac{L}{2}} \in G$, but neither r nor $T_{\frac{L}{2}}$ lies in G. Note also that while some of these groups are isomorphic as "abstract groups" they represent distinct kinds of symmetry groups. In particular there is a finer distinction than isomorphism that reflects the geometric aspects of these groups. This notion will be explored further in the final subsection of this chapter.

10.5 Point Groups of Wallpaper Patterns

In this section we classify the symmetry groups of bounded plane figures and apply the main result to the origin-preserving parts of the isometry groups of wallpaper patterns.

10.5.1 Symmetry Groups of Bounded Plane Figures

We will see that every such symmetry group is isomorphic to a subgroup of $O_2(\mathbb{Z})$. Moreover, if the figure has a "smallest" rotation, then we will see that its symmetry group is either C_n or D_n.

To start, we will be specific about what type of figures we consider. A subset S of $\mathbb{R}^2$ is said to be *bounded* if there is a positive number r such that S is a subset of the closed disk $\{P \in \mathbb{R}^2 : |P| \leq r\}$ of radius r. All examples in the previous section were bounded figures. We begin with a simple but important lemma restricting the type of groups that can arise as the symmetry group of a bounded figure.

Lemma 10.32. *Let S be a bounded figure, and set $G = \text{Sym}(S)$. Then G does not contain a nontrivial translation.*

Proof. Suppose that $\tau \in G$ is a translation, and that τ is translation by the vector b. Then, since G is a group, $\tau^n \in G$ for every positive integer n. Thus, translation by nb is a symmetry of S for every n. Since S is bounded, there is an r such that $|P| \leq r$ for every $P \in S$. But then $\tau^n(P) = P + nb \in S$ for every n. The triangle inequality implies that $|nb| \leq |P| + |P + nb| \leq 2r$. This yields $n\,|b| \leq 2r$. However, since this is true for every n, we must have $|b| = 0$; thus, $b = \mathbf{0}$. Therefore, τ is the identity map. □

To prove the main result about symmetry groups of bounded figures, we need the following facts about compositions of isometries, which we record as a lemma. The proofs are left for exercises.

Lemma 10.33. *The following hold for compositions of isometries:*

1. *The composition of two reflections about lines intersecting at a point P is a rotation about P.*
2. *The composition of two rotations about a common point P is another rotation about P.*
3. *If r and s are rotations about different centers, then $rsr^{-1}s^{-1}$ is a nontrivial translation.*
4. *If r is a rotation about P and f is a reflection not fixing P, then rf is a glide reflection.*
5. *Let r be a rotation and f a reflection in $O_2(R)$. Then $frf = r^{-1}$.*

Proposition 10.34. *Let S be a bounded figure, and set $G = \text{Sym}(S)$. Then G is a subgroup of $O_2(\mathbb{R})$. In particular, every isometry of S is linear.*

Proof. By the lemma there are no nontrivial translations in G. As a consequence, there are no nontrivial glide reflections, since the square of a nontrivial glide is a nontrivial translation. Thus, G must consist solely of rotations and/or reflections. If G is either the trivial group or is generated by a single reflection, then the result is clear. We may then assume that G contains a nontrivial rotation or two reflections. In either case G contains a nontrivial rotation since the composition of two reflections is a rotation. If G contains two rotations r and s, then rs is a rotation if r and s share a common rotation center. However, if they have different centers, then $rsr^{-1}s^{-1}$ is a nontrivial translation. By the lemma, this cannot happen. Therefore, every rotation in G has the same center, which we will view as the origin. If $r \in G$ and if $f \in G$ is a reflection, then fr is a glide reflection if f does preserve the rotation center of r. Since G does not contain a glide reflection, f must preserve the origin. Therefore, all symmetries of S preserve the origin, and so $G \subseteq \mathrm{O}_2(R)$ is comprised of linear isometries. □

Let S be a bounded figure. By the proposition above, $G = \mathrm{Sym}(S)$ is a subgroup of $\mathrm{O}_2(\mathbb{R})$. We can then consider the subgroup R of rotations of S. Of importance to our application is the case where R contains a nontrivial rotation of smallest possible angle. This is the case for instance if S is a regular polygon, but not if S is a circle. The next lemma is a generalization of Exercise 14 of Sect. 10.2.

Lemma 10.35. *Let G be a subgroup of $\mathrm{O}_2(\mathbb{R})$ and set $N = \{g \in G : \det(g) = 1\}$. Then N is a normal subgroup of G with $[G : N] \leq 2$.*

Proof. The determinant function $\det : \mathrm{GL}_n(\mathbb{R}) \to \mathbb{R}^*$ restricts to a homomorphism $\det_G : G \to \{1, -1\}$ with kernel N. Since $G/N = G/\ker(\det_G) \cong \mathrm{im}(\det_G)$ there are two possibilities: Either $\det_G$ is onto, in which case $[G : N] = |G/N| = 2$, or $\mathrm{im}(\det_G) = \{1\}$, i.e., $N = G$ and $[G : N] = 1$. □

Corollary 10.36. *If S is a bounded figure such that $\mathrm{Sym}(S)$ contains a rotation of smallest possible angle, then $\mathrm{Sym}(S)$ is either isomorphic to C_n or to D_n for some n.*

Proof. Let r be a rotation in $G = \mathrm{Sym}(S)$ of smallest possible nonzero angle θ. Recall that θ is only unique up to multiples of 2π. We may assume that $0 < \theta \leq \pi$, since if G contains a rotation by θ, then it contains a rotation by $-\theta$, and if $\pi < \theta < 2\pi$, then $0 < 2\pi - \theta < \pi$, and r^{-1} is rotation by $2\pi - \theta$. If s is another rotation in G, and if s is rotation by ϕ, then there is an integer m with $m\theta \leq \phi < (m+1)\theta$. The symmetry sr^{-m} has rotation angle $\phi - m\theta$, which by the inequality is smaller than θ. By choice of θ, we must have $\phi - m\theta = 0$, so $\phi = m\theta$. Thus, $s = r^m$. Therefore, the group R of rotations of S is the cyclic group generated by r. Moreover, we claim that $\theta = 2\pi/n$ for some n. To prove this, let n be the smallest integer with $2\pi/\theta \leq n$. Then $n - 1 < 2\pi/\theta$. Multiplying these inequalities by θ and rewriting a little gives

$$2\pi \leq n\theta < 2 + \theta,$$

so $0 \leq n\theta - 2\pi < \theta$. However, the rotation angle in the interval $[0, 2\pi]$ for r^n is $n\theta - 2\pi$. Since $r^n \in G$, minimality of θ forces $n\theta - 2\pi = 0$, giving $\theta = 2\pi/n$, as desired. Since we have shown that $R = \langle r \rangle$, and $r^n = \mathrm{id}$ but $r^m \neq \mathrm{id}$ for $0 < m < n$, we see that $|R| = n$, and so $|R| < \infty$. If $G = R$, then $G = C_n$. If not, then G contains a reflection f, so that $frf = r^{-1}$ by Lemma 10.33 (5), and $[G : R] = 2$, by Lemma 10.35. Finally, from Theorem 10.28, with $n = |R|$, we see that G is isomorphic to C_n or D_n. □

Exercises

1. Prove that the composition of two reflections about lines intersecting at a point P is a rotation about P. Moreover, if θ is the angle between the two reflection lines, prove that the composition is a rotation by $\theta/2$.
2. Prove that the composition of two rotations about a common point P is another rotation about P. Moreover, if the rotations are by angles θ and ϕ, prove that the composition is rotation by $\theta + \phi$.

3. If r and s are rotations about different centers, prove that $rsr^{-1}s^{-1}$ is a nontrivial translation.
4. If r is a rotation about P and f is a reflection not fixing P, prove that rf is a glide reflection.
5. Prove that every subgroup of a cyclic group is cyclic.
6. Prove that every subgroup of a dihedral group is either cyclic or dihedral.

10.5.2 Point Groups of Wallpaper Patterns

Similar to the existence of only seven frieze groups, there are only 17 distinct groups of isometries of wallpaper patterns, or two-dimensional tilings of $\mathbb{R}^2$ made by composing certain linear isometries with certain translations. As in the case for bounded plane figures, the "linear parts" are necessarily either cyclic or dihedral, and there are only ten possibilities. These ten finite groups are called the point groups of wallpaper patterns. Translations in this case can be in two independent directions. Once the point groups are classified, it is possible to argue, in analogy with the frieze groups, that there are only 17 ways to combine them with translations to obtain symmetry groups of wallpaper patterns. If one considers crystals, or three-dimensional tilings of $\mathbb{R}^3$, then one finds exactly 230 distinct symmetries. The technicalities of a full treatment of two- and three-dimensional tilings are beyond the scope of this text and the reader is referred elsewhere for detailed treatments of these topics. The remainder of this chapter is dedicated to the determination of the ten point groups and five groups of translations of wallpaper patterns.

To make things more precise, by a wallpaper pattern $\mathcal{W}$ is meant the repetition of a fixed parallelogram $\mathcal{P}$ in the plane $\mathbb{R}^2$ containing a bounded figure. Each pair of nonparallel edges of $\mathcal{P}$ determines two linearly independent displacement vectors v, w and the set of their integer linear combinations $\mathcal{L}_{v,w} := \{mv + nw : m, n \in \mathbb{Z}\}$ is called the integer lattice generated by v and w. The terminal points of the vectors in $\mathcal{L}_{v,w}$ are the vertices of the (two-dimensional) translates of the parallelogram $\mathcal{P}$. It is clear that as a subset of the vector space $\mathbb{R}^2$ the lattice $\mathcal{L}_{v,w}$ is a subgroup under vector addition.

Let $\mathcal{W}$ be a wallpaper pattern with lattice $\mathcal{L}_{v,w}$. Because $\mathcal{W}$ is a plane figure, $\mathrm{Sym}(\mathcal{W})$ is a subgroup of $\mathrm{Isom}(\mathbb{R}^2)$ whose translations are given by $\{t_u : u \in \mathcal{L}_{v,w}\}$. In particular $T \cap \mathrm{Sym}(\mathcal{W}) \cong \mathcal{L}_{v,w}$. Recall that $\mathrm{Isom}(\mathbb{R}^2)$ is the semidirect product of the normal subgroup T by the subgroup $P \cong \mathrm{O}_2(\mathbb{R})$ and that the elements of $\mathrm{Isom}(\mathbb{R}^2)$ are uniquely represented as $f_A \circ t_a$ with $A \in \mathrm{O}_2(\mathbb{R})$ and $a \in \mathbb{R}^2$. We use the identification of the group $\mathrm{Isom}(\mathbb{R}^2)$ with the group structure on $\mathrm{O}_2(\mathbb{R}) \times \mathbb{R}^2$ having as product and inversion:

$$
\begin{aligned}
(A,a) \cdot (B,b) &= (AB, B^{-1}a + b) \\
(A,a)^{-1} &= (A^{-1}, -Aa).
\end{aligned}
$$

Definition 10.37. Given a wallpaper pattern $\mathcal{W}$ its point group is $G_0 := \{A \in \mathrm{O}_2(\mathbb{R}) : (A,a) \in \mathrm{Sym}(\mathcal{W})$ for some $a \in \mathbb{R}^2\}$.

Caution: Just as for frieze Group 6, where we saw that $r \circ T_{\frac{L}{2}} \in G$, but neither r nor $T_{\frac{L}{2}}$ lies in G, we cannot expect that $(I,a) \in \mathrm{Sym}(\mathcal{W})$ if $(A,a) \in \mathrm{Sym}(\mathcal{W})$ for $a \in \mathbb{R}^2$ but $a \notin \mathcal{L}_{v,w}$. This means that $\mathrm{Sym}(\mathcal{W})$ need not be isomorphic to a semidirect product of $\mathcal{L}_{v,w}$ by G_0. For an example of a wallpaper pattern of this type see the example labeled $p4m$ at the end of the chapter. The point group however is the image of $\mathrm{Sym}(\mathcal{W})$ under a surjective group homomorphism.

Proposition 10.38. *Let $\mathcal{W}$ be a wallpaper pattern with lattice $\mathcal{L}_{v,w}$ and point group G_0. The mapping*

$$\begin{aligned} \pi : \mathrm{Sym}(\mathcal{W}) &\to G_0 \\ : (A, a) &\mapsto A \end{aligned}$$

is a surjective group homomorphism with kernel naturally isomorphic to $\mathcal{L}_{v,w}$.

Proof. That π is a homomorphism follows directly from the interpretation above of the group operation in $\mathrm{Sym}(\mathcal{W})$. The definition of G_0 shows that π is surjective and, with I denoting the identity matrix, the kernel of π is $\{(I, a) : t_a \in T \cap \mathrm{Sym}(\mathcal{W})\}$. But $t_a \in T \cap \mathrm{Sym}(\mathcal{W})$ implies that $a \in \mathcal{L}_{v,w}$. □

That the point group is either cyclic or dihedral will follow from realizing that it is isomorphic to the group of symmetries of a bounded figure in the plane.

Lemma 10.39. *G_0 is isomorphic to a subgroup of $O_2(\mathbb{R})$.*

Proof. To show that G_0 is isomorphic to a subgroup of $O_2(\mathbb{R})$ amounts to showing that if (A, a) and $(B, b) \in \mathrm{Sym}(\mathcal{W})$ then there are $c, d \in T$ with $(AB, c) \in \mathrm{Sym}(\mathcal{W})$ and $(A^{-1}, d) \in \mathrm{Sym}(\mathcal{W})$ where AB is the product of the orthogonal matrices A and B, and A^{-1} is the matrix inverse of A. These conditions follow from the expressions for products and inversion above in $\mathrm{Isom}(\mathbb{R}^2)$ with respect to the given representation of its elements and the fact that $\mathrm{Sym}(\mathcal{W})$ is a subgroup of $\mathrm{Isom}(\mathbb{R}^2)$. □

While the matrix representations for elements of G_0 with respect to the standard basis for $\mathbb{R}^2$ lie in $O_2(\mathbb{R})$, the matrices with respect to a lattice basis can only be expected to lie in $GL_2(\mathbb{Z})$.

Lemma 10.40. *Let $\mathcal{W}$ be a wallpaper pattern with lattice $\mathcal{L}_{v,w}$ and point group G_0. Then G_0 acts on $\mathcal{L}_{v,w}$ in the sense that if $A \in G_0$ and $u \in \mathcal{L}_{v,w}$, then $Au \in \mathcal{L}_{v,w}$. In particular, with respect to the basis $\{v, w\}$ of $\mathbb{R}^2$, the elements of G_0 lie in $GL_2(\mathbb{Z})$.*

Proof. The first assertion follows from the following bit of calculation using the representation above of the product in $\mathrm{Sym}(\mathcal{W})$. Suppose that $A \in G_0$ so that $(A^{-1}, a) \in \mathrm{Sym}(\mathcal{W})$ for some $a \in \mathbb{R}^2$. Since $u \in \mathcal{L}_{v,w}$, we have (I, u) as well as the conjugation $c_{(A^{-1},a)^{-1}}(I, u)$ as elements of $\mathrm{Sym}(\mathcal{W})$. Calculate $c_{(A^{-1},a)^{-1}}(I, u)$ to obtain

$$\begin{aligned} (A^{-1}, a)^{-1}(I, u)(A^{-1}, a) &= (A, -A^{-1}a)(I, u)(A^{-1}, a) \\ &= (A, -A^{-1}a)(A^{-1}, Au + a) \\ &= (I, -a + Au + a) \\ &= (I, Au) \in \mathrm{Sym}(\mathcal{W}), \end{aligned}$$

as desired.

The second assertion follows from the first. Since the elements of G_0 act on $\mathcal{L}_{v,w}$, it must be the case that for $A \in G_0$, the vectors Av and Aw lie in $\mathcal{L}_{v,w}$, hence are integer linear combinations of v and w. The columns of the matrix representation of A with respect to the basis $\{v, w\}$ are exactly the coefficients of v and w in the expansions of Av and Aw and, because A^{-1} acts as the inverse of A, this matrix is necessarily invertible. □

Corollary 10.41. *The point group G_0 of a wallpaper pattern with lattice $\mathcal{L}_{v,w}$ is isomorphic to a subgroup of the symmetry group of a bounded plane figure, hence is isomorphic to C_n or to D_n for some n.*

Proof. Consider a circular disc $\mathcal{D}$ centered at the origin in $\mathbb{R}^2$ that contains the terminal points of the vectors v and w. Then $\mathcal{D} \cap \mathcal{L}_{v,w}$, the collection of lattice points contained in $\mathcal{D}$, is finite and certainly a bounded region. The action of G_0 on $\mathcal{L}_{v,w}$ via isometries induces a homomorphism $\rho : G_0 \to \mathrm{Sym}(\mathcal{D} \cap \mathcal{L}_{v,w})$ simply by restricting the action of $g \in G_0$ to the points of $\mathcal{D} \cap \mathcal{L}_{v,w}$. Note that ρ must be injective. Its kernel consists of all $g \in G_0$ restricting to the identity on $\mathcal{D} \cap \mathcal{L}_{v,w}$. But $\mathcal{D} \cap \mathcal{L}_{v,w}$ contains the vectors v and w, which comprise a basis for $\mathbb{R}^2$, and a linear transformation which acts as the identity on a basis must be the identity map. Thus G_0 is isomorphic to a subgroup of $\mathrm{Sym}(\mathcal{D} \cap \mathcal{L}_{v,w})$, hence to a subgroup of a cyclic group or of a dihedral group. Since every subgroup of such a group is again either cyclic or dihedral (Exercise 6 of Sect. 10.5.1) the result follows. □

We finally come to the classification of the point groups of wallpaper patterns. In the statement of the theorem the cyclic group C_2 is generated by a rotation through 180°. The group D_1, also cyclic of order 2, is generated by a reflection, and D_2, which is isomorphic to the Klein four group $\mathbb{Z}_2 \times \mathbb{Z}_2$, is generated by two reflections.

Theorem 10.42. *Let $\mathcal{W}$ be a wallpaper pattern with lattice $\mathcal{L}_{v,w}$ and point group G_0. Then G_0 is one of the ten groups $C_1, C_2, C_3, C_4, C_6, D_1, D_2, D_3, D_4$, or D_6.*

Remark 10.43. Of these ten groups the only two that are isomorphic as "abstract groups" are the two cyclic groups of order 2, C_2 and D_1. As in the earlier case of distinct frieze groups that are "abstractly" isomorphic, C_2 and D_1 are distinct as point groups of wallpaper patterns. This notion will be made clearer in the next subsection.

Proof of the Theorem. From Corollary 10.36, G_0 is either cyclic or dihedral. If G_0 contains no rotation, then $G_0 \cong C_1$ or $G_0 \cong D_1$, i.e., consists of just the identity or the identity and a reflection. Otherwise $G_0 \cong C_n$ or $G_0 \cong D_n$ with $n > 1$ and contains a rotation r through some minimal angle $\frac{2\pi}{n}$. The matrix for r with respect to the standard basis is then $\begin{pmatrix} \cos\frac{2\pi}{n} & -\sin\frac{2\pi}{n} \\ \sin\frac{2\pi}{n} & \cos\frac{2\pi}{n} \end{pmatrix}$ while the matrix for r with respect to the lattice basis $\{v, w\}$ is in $O_2(\mathbb{Z})$ and these matrices are similar. Since similar matrices have the same trace, it follows that $2\cos\frac{2\pi}{n}$ is an integer, i.e., $\cos\frac{2\pi}{n} \in \{0, \pm 1, \pm\frac{1}{2}\}$ and $n \in \{1, 2, 3, 4, 6\}$. Finally, if G_0 contains no reflection, then $G_0 \cong C_i$ for one of these i. Otherwise $G_0 \cong D_i$. □

10.5.3 Equivalence Versus Isomorphism

Recall that several of the frieze groups, while representing symmetries of distinct frieze patterns, are nevertheless isomorphic as abstract groups. Similarly the point groups C_2 and D_1 are isomorphic as groups but arise from distinct wallpaper patterns. The notion of equivalence within $\mathrm{Isom}(\mathbb{R}^2)$ explains this finer classification. In the following definition we make use of the identification of the lattice associated with a wallpaper pattern $\mathcal{W}$ with the translation subgroup of $\mathrm{Sym}(\mathcal{W})$.

Definition 10.44. Let $\mathcal{W}$ (resp. $\mathcal{W}'$) be a wallpaper pattern with lattice $\mathcal{L}_{v,w}$ (resp. $\mathcal{L}_{v',w'}$). The groups $\mathrm{Sym}(\mathcal{W})$ and $\mathrm{Sym}(\mathcal{W}')$ are said to be equivalent if there is a group isomorphism $\varphi : \mathrm{Sym}(\mathcal{W}) \to \mathrm{Sym}(\mathcal{W}')$ with $\varphi(\mathcal{L}_{v,w}) = \mathcal{L}_{v',w'}$.

Example 10.45. Suppose that $\mathrm{Sym}(\mathcal{W})$ and $\mathrm{Sym}(\mathcal{W}')$ consist only of translations, i.e., the point group just consists of the identity, so that

$$\mathrm{Sym}(\mathcal{W}) = \{(I, mv + nw) : m, n \in \mathbb{Z}\} \cong \mathcal{L}_{v,w}$$

and

$$\mathrm{Sym}(\mathcal{W}') = \{(I, mv' + nw') : m, n \in \mathbb{Z}\} \cong \mathcal{L}'_{v',w'}.$$

The isomorphism $\varphi : \mathrm{Sym}(\mathcal{W}) \to \mathrm{Sym}(\mathcal{W}')$ given by $\varphi((I, mv + nw)) = (I, mv' + nw')$ demonstrates the equivalence.

The presence of a nontrivial point group, however, indicates the restrictive nature of equivalence.

Theorem 10.46. *Let $\mathcal{W}$ and $\mathcal{W}'$ be wallpaper patterns with lattices $\mathcal{L}_{v,w}$ and. $\mathcal{L}_{v',w'}$ and point groups G_0 and G_0', both viewed as subgroups of $\mathrm{O}_2(\mathbb{Z})$. If the groups $\mathrm{Sym}(\mathcal{W})$ and $\mathrm{Sym}(\mathcal{W}')$ are equivalent then there is a matrix $U \in \mathrm{GL}_2(\mathbb{Z})$ with $G_0' = U^{-1}G_0U$.*

In other words, equivalent symmetry groups have point groups G_0' and G_0 which are conjugate as subgroups of $\mathrm{GL}_2(\mathbb{Z})$, i.e., the elements of G_0' are simultaneously similar to those in G_0 by the same integer matrix U.

Example 10.47. The symmetry groups of wallpaper patterns with $G_0 = C_2$ represented as $\left\{I, \begin{pmatrix} -1 & 0 \\ 0 & -1 \end{pmatrix}\right\}$ and $G_0' = D_1$ represented as $\left\{I, \begin{pmatrix} 0 & 1 \\ 1 & 0 \end{pmatrix}\right\}$ are not equivalent. Recall that similar matrices have the same determinant. Since both matrices in G_0 have determinant equal to 1 while the nonidentity element in G_0' has determinant equal to -1, these groups cannot be equivalent.

Proof of the Theorem. Let $\varphi : \mathrm{Sym}(\mathcal{W}) \to \mathrm{Sym}(\mathcal{W}')$ be a group isomorphism satisfying $\varphi(\mathcal{L}_{v,w}) = \mathcal{L}_{v',w'}$. Define U to be the matrix $\begin{pmatrix} a & c \\ b & d \end{pmatrix}$ where

$$\varphi(v) = av' + bw'$$
$$\varphi(w) = cv' + dw'.$$

Since $\varphi(v), \varphi(w) \in \mathcal{L}_{v',w'}$, which consists of all integer linear combinations of v' and w' and the expansion of these vectors in the basis $\{v', w'\}$ is unique, it must be the case that $a, b, c, d \in \mathbb{Z}$. But the rank of U must be equal to 2 because $\varphi(\mathcal{L}_{v,w}) = \mathcal{L}_{v',w'}$, i.e., the image of φ contains the linearly independent vectors v', w'. Indeed for some integers a', b', c', d',

$$v' = a'\varphi(v) + b'\varphi(w) = a'(av' + bw') + b'(cv' + dw')$$
$$w' = c'\varphi(v) + d'\varphi(w) = c'(av' + bw') + d'(cv' + dw').$$

Again by the uniqueness of representations in the basis $\{v', w'\}$ it follows that

$$\begin{pmatrix} a & c \\ b & d \end{pmatrix}\begin{pmatrix} a' & c' \\ b' & d' \end{pmatrix} = I.$$

Thus U is invertible and its inverse $\begin{pmatrix} a' & c' \\ b' & d' \end{pmatrix}$ is also an integer matrix. Writing $(I, u) \in \mathrm{Sym}(\mathcal{W})$ for the element corresponding to $u \in \mathcal{L}_{v,w}$, we have shown $\varphi(I, u) = (I, Uu)$.

Next let $(A, a) \in \mathrm{Sym}(\mathcal{W})$ and set $(A', a') = \varphi(A, a)$. From the formula for the product in $\mathrm{Sym}(\mathcal{W})$, in particular the action of G_0 on T:

$$(A, a)^{-1}(I, u)(A, a) = (I, AU),$$

and the fact that φ is a group homomorphism, we have for $u \in \mathcal{L}_{v,w}$,

$$\begin{aligned}(I, UAu) &= \varphi(I, Au) \\ &= \varphi((A, a)^{-1}(I, u)(A, a)) \\ &= (A', a')^{-1}(I, Uu)(A', a') \\ &= (I, A'Uu)\end{aligned}$$

so that $UAu = A'Uu$ for every lattice vector u. Since the lattice spans $\mathbb{R}^2$ it must be the case that the matrices satisfy $UA = A'U$, i.e., that the matrices in G_0' are similar via U to those in G_0. □

Remark 10.48. The inequivalence of C_2 and D_1 is therefore a consequence of the different actions of these two cyclic groups of order 2 on the respective lattices.

Exercises

1. Find an explicit isomorphism between D_2 and $\mathbb{Z}_2 \times \mathbb{Z}_2$.
2. Determine the elements of $O_2(\mathbb{Z})$ and find the point group to which it is isomorphic.
3. Find an infinite abelian subgroup of $O_2(\mathbb{R})$.
4. Find an infinite cyclic subgroup of $O_2(\mathbb{R})$.

10.5.4 The Five Lattice Types

For the remainder we will use the symbol T both for the translation group of a wallpaper pattern and for its associated lattice. In the previous section we proved that the point group G_0 of a wallpaper pattern is isomorphic to one of the ten groups $\{C_n, D_n : n = 1, 2, 3, 4, 6\}$ and that this set has only nine non-isomorphic groups. Moreover, while $C_2 \cong D_1$ as abstract groups, they are distinguished by their actions on T. By fixing a basis $\{t_1, t_2\}$ of T, we have an isomorphism $T \cong \mathbb{Z}^2$, and using the basis, the action of G_0 on T induces a group homomorphism $G_0 \to \mathrm{Aut}(\mathbb{Z}^2) \cong \mathrm{GL}_2(\mathbb{Z})$. In other words, a choice of basis together with the action of G_0 on T gives us a representation of G_0 as a specific subgroup of $\mathrm{GL}_2(\mathbb{Z})$.

Viewing lattices geometrically, we will see that there are five types of lattices with respect to the G_0-action; parallelogram, rectangular, square, rhombic, and hexagonal. We will be specific in what we mean as we look at the actions of the ten groups above on T.

10.5.4.1 G_0 Is One of C_1 or C_2: Parallelogram Lattices

As mentioned above, we will represent a point group as a subgroup of $O_2(\mathbb{Z})$ by choosing a basis $\{t_1, t_2\}$ for T. The groups C_1 and C_2 are very easy to describe and their description does not depend on the basis. If $G_0 = C_1$, then

$$C_1 = \left\{ \begin{pmatrix} 1 & 0 \\ 0 & 1 \end{pmatrix} \right\}.$$

On the other hand, if $G_0 = C_2$, then the rotation of 180° is multiplication by -1 on T. Therefore,

$$C_2 = \left\langle \begin{pmatrix} -1 & 0 \\ 0 & -1 \end{pmatrix} \right\rangle.$$

The lattice in these cases is called a *parallelogram lattice*.

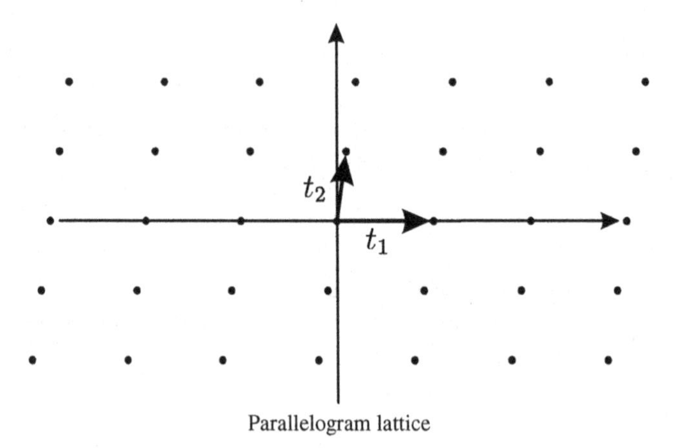

Parallelogram lattice

10.5.4.2 G_0 Is One of C_n or D_n for $n \geq 3$

The following lemma will help us find a convenient basis for T in these cases.

Lemma 10.49. *Suppose that G_0 contains a rotation r about an angle $2\pi/n$ for $n \geq 3$. If t is a nonzero element of T of minimal length, then $\{t, r(t)\}$ is a basis for T (i.e., every lattice vector in T is an integer linear combination of t and $r(t)$).*

Proof. Let $\{t_1, t_2\}$ be a basis for T. Then

$$t = at_1 + bt_2$$
$$r(t) = ct_1 + dt_2$$

for some integers a, b, c, d. The set $\{t, r(t)\}$ is linearly independent because $n > 2$, so we can solve for t_1 in the two equations above: $t_1 = \alpha t + \beta r(t)$ for some rational numbers α, β. We claim that $\alpha, \beta \in \mathbb{Z}$. Write $\alpha = \alpha_0 + \varepsilon$ and $\beta = \beta_0 + \varepsilon'$ with $\alpha_0, \beta_0 \in \mathbb{Z}$ and $|\varepsilon|, |\varepsilon'| \leq 1/2$. We have $s = \alpha_0 t + \beta_0 r(t) \in T$, so $(t_1 - s) = \varepsilon t + \varepsilon' r(t) \in T$. Since t and $r(t)$ are not parallel, we see that

$$\|t_1 - s\| = \left\|\varepsilon t + \varepsilon' r(t)\right\| < \|\varepsilon t\| + \left\|\varepsilon' r(t)\right\| \leq \frac{1}{2}\left(\|t\| + \|r(t)\|\right).$$

Thus $||t_1 - s|| < ||t||$, a contradiction to the minimality of $\|t\|$, unless $s = t_1$ (i.e., $\varepsilon = \varepsilon' = 0$). Therefore, $t_1 = s$ is a $\mathbb{Z}$-linear combination of t and $r(t)$. Similarly, t_2 is a $\mathbb{Z}$-linear combination of t and $r(t)$. Since $\{t_1, t_2\}$ is a basis of T, the set $\{t, r(t)\}$ is also a basis for T. □

10.5.4.3 G_0 Is One of C_4, D_4: Square Lattices

Let r be a rotation by 90°. By Lemma 10.49, if $t = t_1$ is a vector in T of minimal length, then $\{t_1, r(t_1)\}$ is a basis for T. The lattice is called a *square lattice*.

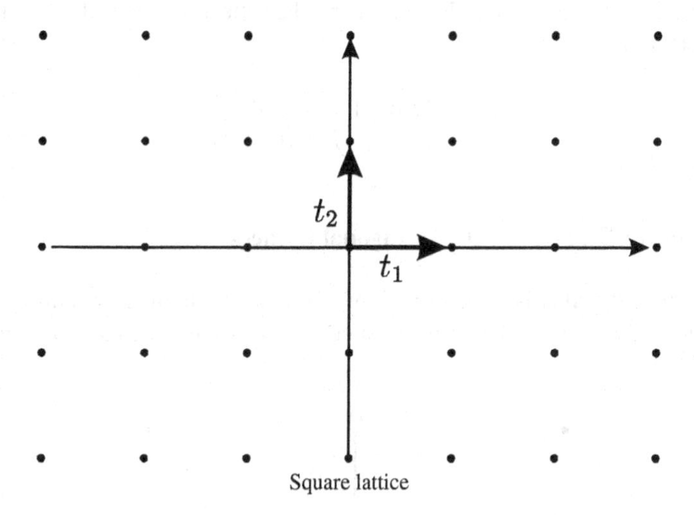

Square lattice

With respect to this basis, we see that if $G_0 = C_4 = \langle r \rangle$, then the representation of G_0 by this basis is

$$C_4 = \left\langle \begin{pmatrix} 0 & -1 \\ 1 & 0 \end{pmatrix} \right\rangle.$$

On the other hand, if $G_0 = D_4$, then G_0 contains a reflection f. The four elements $f, rf, r^2 f, r^3 f$ are all the reflections in G_0. These reflections must preserve the set of vectors in T of minimal length; four such vectors are $\pm t_1, \pm t_2$. However, a short argument shows that any other point on the circle of radius $\|t_1\|$ centered at the origin is a distance of less than $\|t_1\|$ from one of these four points. The figure below makes this easy to see. The difference of these two vectors would then be a vector in T of length less than $\|t_1\|$. Since this is impossible, we see that the four vectors above are all the vectors of minimal length in T. The four lines of reflection are then given in the following picture.

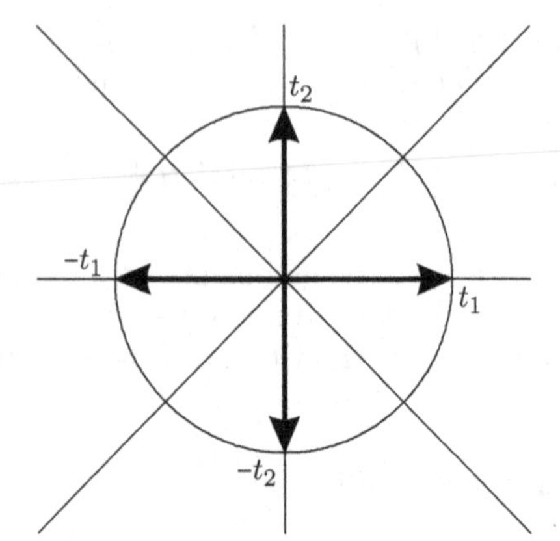

The vectors of minimal length in T when $G_0 = D_4$

Since D_4 is generated by r and any reflection, using the reflection about the line parallel to t_1, we obtain the representation

$$D_4 = \left\langle \begin{pmatrix} 0 & -1 \\ 1 & 0 \end{pmatrix}, \begin{pmatrix} 1 & 0 \\ 0 & -1 \end{pmatrix} \right\rangle.$$

10.5.4.4 G_0 Is One of C_3, D_3, C_6, D_6: Hexagonal Lattices

Let r be a rotation by 120°. If t_1 is a vector in T of minimal length, then by setting $t_2 = r(t_1)$, the set $\{t_1, t_2\}$ is a basis for T, by Lemma 10.49. The lattice in this case is called a *hexagonal lattice*.

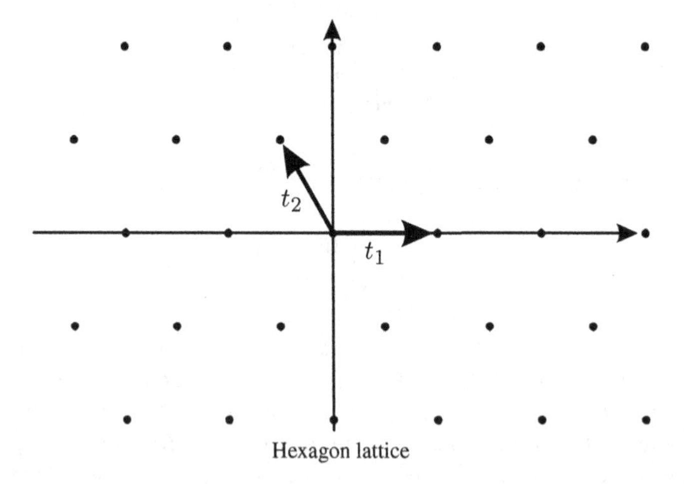

Hexagon lattice

The group C_3 is generated by r and C_6 is generated by a 60° rotation; thus, we obtain

$$C_3 = \left\langle \begin{pmatrix} 0 & -1 \\ 1 & -1 \end{pmatrix} \right\rangle$$

and

$$C_6 = \left\langle \begin{pmatrix} 1 & -1 \\ 1 & 0 \end{pmatrix} \right\rangle.$$

The figure below indicates that we have six vectors in T of minimal length.

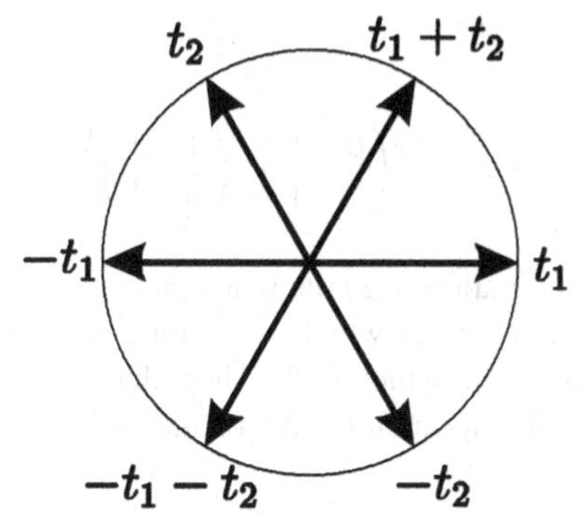

The vectors of minimal length when $G_0 = D_6$

Any point on the circle above other than the six shown is a distance less than $\|t_1\|$ from one of these six points. This shows that these six vectors are all the vectors of minimal length in T.

If $G_0 = D_3$ or D_6, then G_0 contains three or six reflections, respectively. Any reflection must permute the six vectors in the previous figure. For $G_0 = D_6$, then we see six lines of reflection in the following diagram.

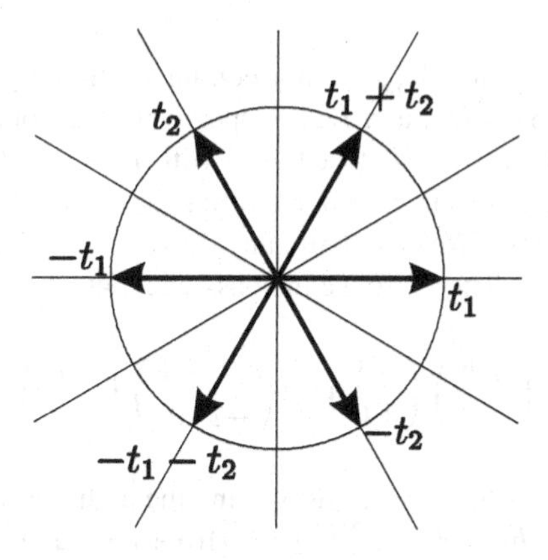

The group D_6 is generated by C_6 and any reflection; using the reflection that fixes t_1, we have

$$D_6 = \left\langle \begin{pmatrix} 1 & -1 \\ 1 & 0 \end{pmatrix}, \begin{pmatrix} 1 & -1 \\ 0 & -1 \end{pmatrix} \right\rangle.$$

If $G_0 = D_3$, then the point group contains three reflections. The lines of reflection are separated by $60°$ angles; if f is a reflection in D_3, then rf is a reflection whose line of reflection makes a $60°$ angle with that of f. The reflection lines for D_3 must be reflection lines for D_6 since D_3 is a subgroup of D_6. We then have two possibilities: The three lines are the lines that are at angles $30°, 90°, 150°$ with

t_1 or are the lines at angles $0°, 60°, 120°$ with t_1. This says that D_3 can act in two ways with respect to this basis. We write $D_{3,l}$ and $D_{3,s}$ to distinguish these two actions; therefore, generating $D_{3,l}$ and $D_{3,s}$ with the 120° rotation and with the reflection about the 30° and the 0° reflection lines, respectively, we have

$$D_{3,l} = \left\langle \begin{pmatrix} 0 & -1 \\ 1 & -1 \end{pmatrix}, \begin{pmatrix} 1 & 0 \\ 1 & -1 \end{pmatrix} \right\rangle$$

and

$$D_{3,s} = \left\langle \begin{pmatrix} 0 & -1 \\ 1 & -1 \end{pmatrix}, \begin{pmatrix} 1 & -1 \\ 0 & -1 \end{pmatrix} \right\rangle.$$

To give meaning to this subscript notation, we note that l and s stand for long and short, respectively. The vectors t_1 and t_2 span a parallelogram which has a long and a short diagonal. The group $D_{3,s}$ contains a reflection about the 60° line, which is the short diagonal. The group $D_{3,l}$ has a reflection across the 150° line, which is parallel to the long diagonal.

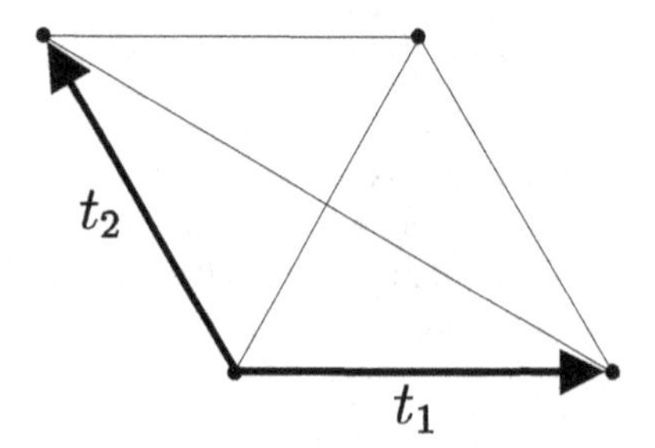

We show that the groups $D_{3,l}$ and $D_{3,s}$ are not conjugate in $\mathrm{GL}_2(\mathbb{Z})$. This will tell us that two wallpaper groups with point groups $D_{3,l}$ and $D_{3,s}$, respectively, are not equivalent, by Theorem 10.46. To prove this, suppose there is a matrix $U \in \mathrm{GL}_2(\mathbb{Z})$ with $D_{3,l} = UD_{3,s}U^{-1}$. Because conjugation preserves determinants and the determinant of a reflection is -1, the three reflections of $D_{3,s}$ must be sent to the three reflections of $D_{3,l}$. We can obtain any reflection (in D_3) from any other reflection by conjugation by one of I, r, or r^2. Therefore we may assume that

$$\begin{pmatrix} a & b \\ c & d \end{pmatrix} \begin{pmatrix} 0 & 1 \\ 1 & 0 \end{pmatrix} = \begin{pmatrix} 0 & -1 \\ -1 & 0 \end{pmatrix} \begin{pmatrix} a & b \\ c & d \end{pmatrix}$$

for some $a, b, c, d \in \mathbb{Z}$ with $ad - bc = \pm 1$. Multiplying the matrices and simplifying yields $d = -a$ and $c = -b$. Since $\pm 1 = ad - bc = b^2 - a^2 = (b-a)(b+a)$ is a factorization in integers, one term is 1 and the other is -1, yielding four cases, $a = \pm 1$ and $b = 0$ or $a = 0$ and $b = \pm 1$. Conjugation by $-I_2$ is the identity; therefore, we may assume that

$$\begin{pmatrix} a & b \\ c & d \end{pmatrix} = \begin{pmatrix} 1 & 0 \\ 0 & -1 \end{pmatrix}$$

or

$$\begin{pmatrix} a & b \\ c & d \end{pmatrix} = \begin{pmatrix} 0 & 1 \\ -1 & 0 \end{pmatrix}.$$

However, since

$$\begin{pmatrix} 1 & 0 \\ 0 & -1 \end{pmatrix} \begin{pmatrix} 1 & -1 \\ 0 & -1 \end{pmatrix} \begin{pmatrix} 1 & 0 \\ 0 & -1 \end{pmatrix}^{-1} = \begin{pmatrix} 1 & 1 \\ 0 & -1 \end{pmatrix}$$

and

$$\begin{pmatrix} 0 & 1 \\ -1 & 0 \end{pmatrix} \begin{pmatrix} 1 & -1 \\ 0 & -1 \end{pmatrix} \begin{pmatrix} 0 & 1 \\ -1 & 0 \end{pmatrix}^{-1} = \begin{pmatrix} -1 & 0 \\ 1 & 1 \end{pmatrix},$$

neither conjugation sends $D_{3,s}$ to $D_{3,l}$ since neither of these results is an element of $D_{3,l}$. The groups $D_{3,l}$ and $D_{3,s}$ are thus not conjugate in $\mathrm{GL}_2(\mathbb{Z})$.

10.5.4.5 G_0 Is One of D_1, D_2: Rectangular or Rhombic Lattices

If $G_0 = D_1$ or D_2, then G_0 does not contain a rotation of order at least 3. Therefore, we cannot apply Lemma 10.49 to obtain a basis for T. We produce a basis in another way. In each of these cases we have a nontrivial reflection f in G_0. Let $t \in T$ be a nonzero vector not parallel to the line of reflection of f. Since f maps T to T, the vectors $t + f(t)$ and $t - f(t)$ are elements of T, so T contains nonzero vectors both parallel and perpendicular to the line of reflection.

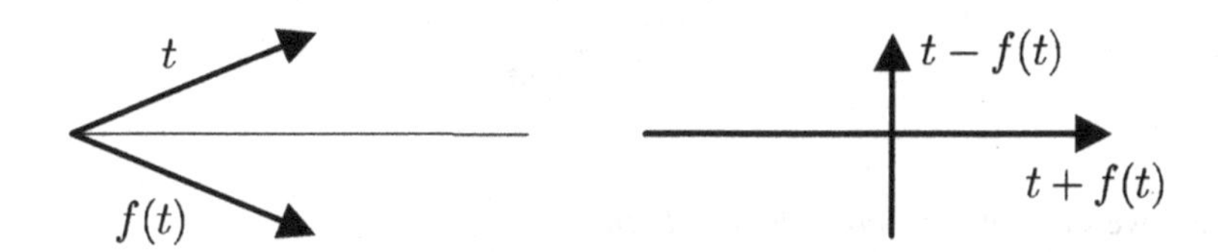

Let s_1 and s_2 be nonzero vectors of minimal length parallel and perpendicular, respectively, to the reflection line. The discrete nature of T implies that such vectors exist, and that any vector parallel to (resp. perpendicular to) this line is an integer multiple of s_1 (resp. s_2). Therefore, for any $t \in T$, we have

$$t + f(t) = m_t s_1,$$
$$t - f(t) = n_t s_2$$

for some $m_t, n_t \in \mathbb{Z}$. Solving for t gives

$$t = \frac{m_t}{2} s_1 + \frac{n_t}{2} s_2.$$

If, for every $t \in T$, both integers m_t, n_t are even, the set $\{s_1, s_2\}$ spans T, and so is a basis for T. On the other hand, if m_t or n_t is odd for some t, then both have to be odd, else $\frac{1}{2}s_1$ or $\frac{1}{2}s_2$ is in T, a contradiction. If we set $t_1 = \frac{1}{2}(s_1 + s_2)$ and $t_2 = \frac{1}{2}(s_1 - s_2) = f(t_1)$, then $t_1, t_2 \in T$, and

$$
\begin{aligned}
t &= \frac{m_t}{2}s_1 + \frac{n_t}{2}s_2 = \left(\frac{m_t + n_t}{2}\right)\left(\frac{s_1 + s_2}{2}\right) + \left(\frac{m_t - n_t}{2}\right)\left(\frac{s_1 - s_2}{2}\right) \\
&= m'_t t_1 + n'_t t_2
\end{aligned}
$$

with $m'_t, n'_t \in \mathbb{Z}$. Since any t is then an integral linear combination of t_1 and t_2, the set $\{t_1, t_2\}$ is a basis for T.

To summarize these two cases, we either have a basis $\{t_1, t_2\}$ of two orthogonal vectors, one of which is fixed by a reflection in G_0,

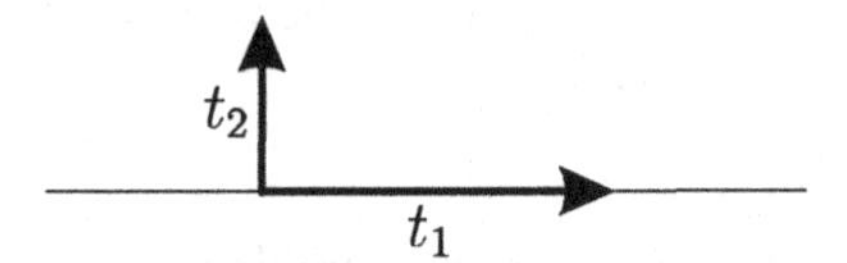

or we have a basis of vectors of the same length with a reflection that interchanges them.

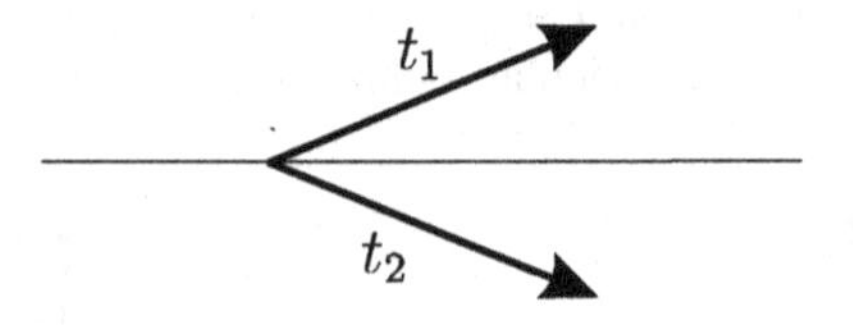

In the first case we say that T is a *rectangular lattice*

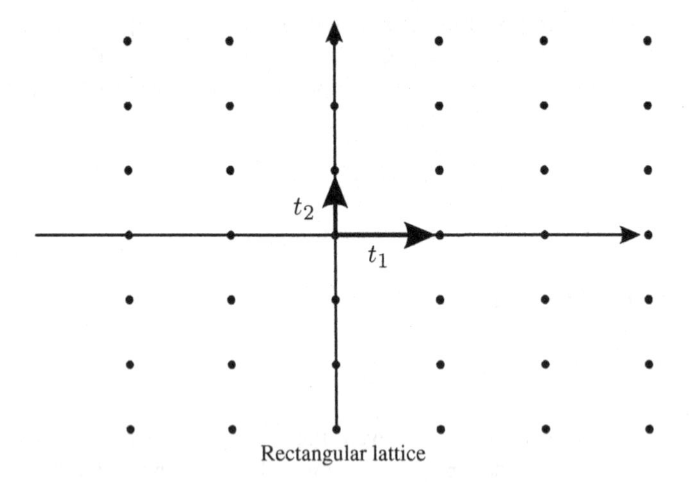

Rectangular lattice

and in the second case that T is a *rhombic lattice*.

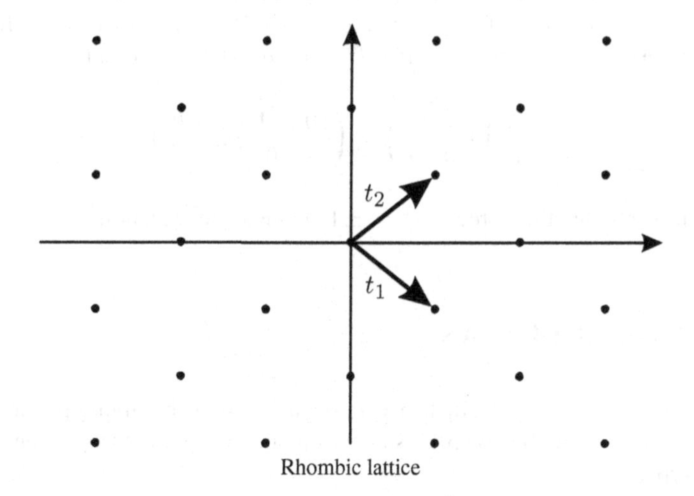

Rhombic lattice

We can now get matrix representations for D_1 and D_2. For each group there are two possibilities, corresponding to two different actions on T. We subscript the group by p for rectangular and c for rhombic to match the notation used for wallpaper groups that is standard in the literature. We have

$$D_{1,p} = \left\langle \begin{pmatrix} 1 & 0 \\ 0 & -1 \end{pmatrix} \right\rangle$$

and

$$D_{1,c} = \left\langle \begin{pmatrix} 0 & 1 \\ 1 & 0 \end{pmatrix} \right\rangle,$$

while for D_2, which contains a rotation of 180°, we obtain

$$D_{2,p} = \left\langle \begin{pmatrix} -1 & 0 \\ 0 & -1 \end{pmatrix}, \begin{pmatrix} 1 & 0 \\ 0 & -1 \end{pmatrix} \right\rangle$$

and

$$D_{2,c} = \left\langle \begin{pmatrix} -1 & 0 \\ 0 & -1 \end{pmatrix}, \begin{pmatrix} 0 & 1 \\ 1 & 0 \end{pmatrix} \right\rangle.$$

We prove that $D_{1,p}$ and $D_{1,c}$ are not conjugate in $\mathrm{GL}_2(\mathbb{Z})$, nor are $D_{2,p}$ and $D_{2,c}$. This will show that no wallpaper group whose point group is one of these is equivalent to a wallpaper group whose point group is another. For $D_{1,p}$ and $D_{1,c}$, suppose that

$$\begin{pmatrix} a & b \\ c & d \end{pmatrix} \begin{pmatrix} 1 & 0 \\ 0 & -1 \end{pmatrix} = \begin{pmatrix} 0 & 1 \\ 1 & 0 \end{pmatrix} \begin{pmatrix} a & b \\ c & d \end{pmatrix}$$

for some $a, b, c, d \in \mathbb{Z}$ with $ad - bc = \pm 1$. Multiplying these and setting the two sides equal yields $d = -b$ and $c = -a$. Then $ad - bc = -2ab$, which is not ± 1 since a and b are integers. Therefore, $D_{1,p}$ and $D_{1,c}$ are not conjugate in $\mathrm{GL}_2(\mathbb{Z})$. For $D_{2,p}$ and $D_{2,c}$, the previous calculation shows that we need only check that there are no $a, b, c, d \in \mathbb{Z}$ with $ad - bc = \pm 1$ and

$$\begin{pmatrix} a & b \\ c & d \end{pmatrix}\begin{pmatrix} 1 & 0 \\ 0 & -1 \end{pmatrix} = \begin{pmatrix} 0 & -1 \\ -1 & 0 \end{pmatrix}\begin{pmatrix} a & b \\ c & d \end{pmatrix}.$$

Similar calculations show that this forces $2ab = \pm 1$, again a contradiction.

10.6 The 17 Wallpaper Groups

We summarize the classification of wallpaper patterns and groups by giving pictures for each of the 17 different patterns. The table below gives standard names for these groups along with their point group and lattice type.

Standard name	Point group	Lattice type
p1	C_1	Parallelogram
p2	C_2	Parallelogram
pm	D_1	Rectangular
pg	D_1	Rectangular
cm	D_1	Rhombic
pmm	D_2	Rectangular
pmg	D_2	Rectangular
pgg	D_2	Rectangular
cmm	D_2	Rhombic
p3	C_3	Hexagonal
p3m1	D_3	Hexagonal
p31m	D_3	Hexagonal
p4	C_4	Square
p4m	D_4	Square
p4g	D_4	Square
p6	C_6	Hexagonal
p6m	D_6	Hexagonal

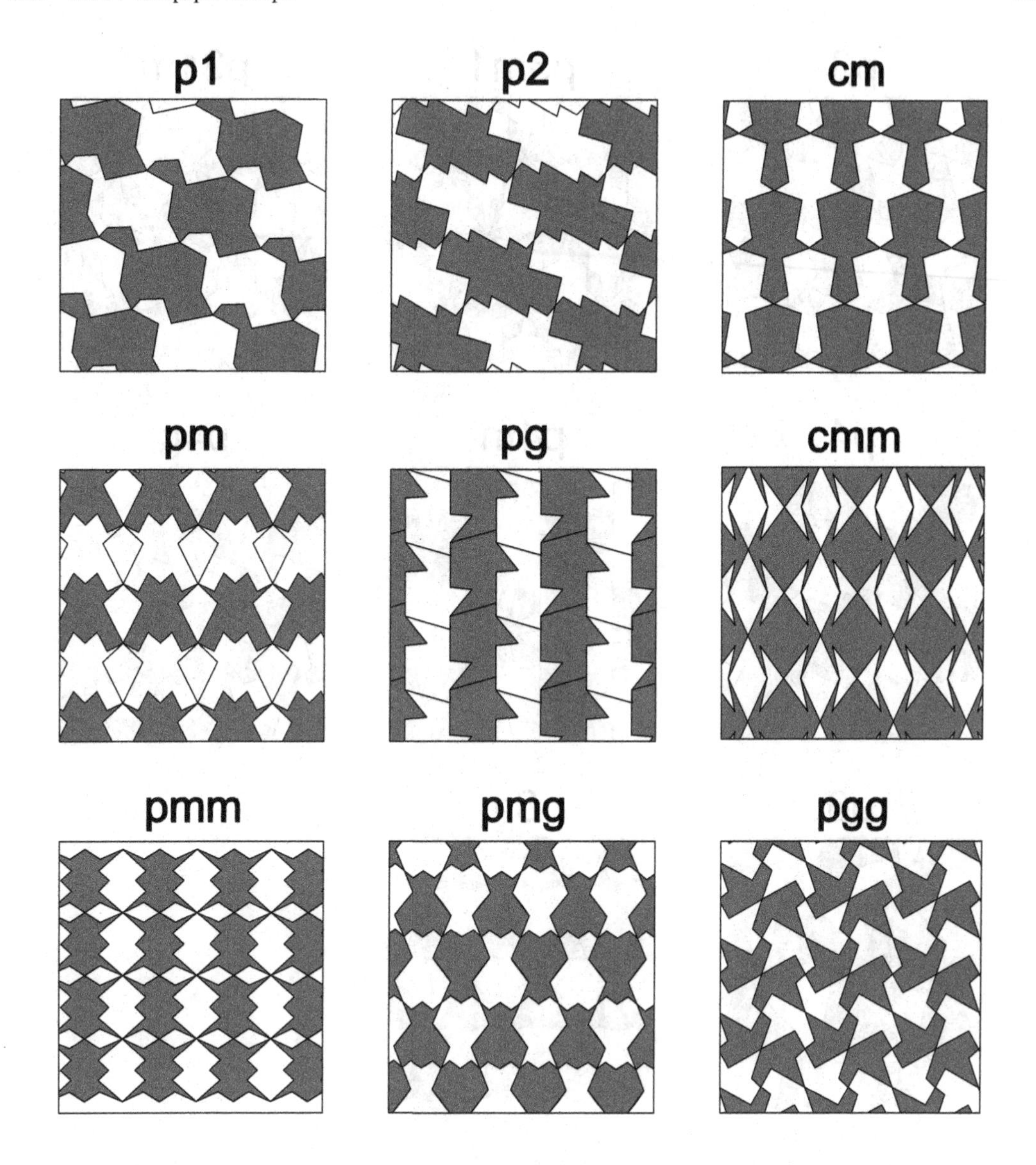
p1
p2
cm
pm
pg
cmm
pmm
pmg
pgg

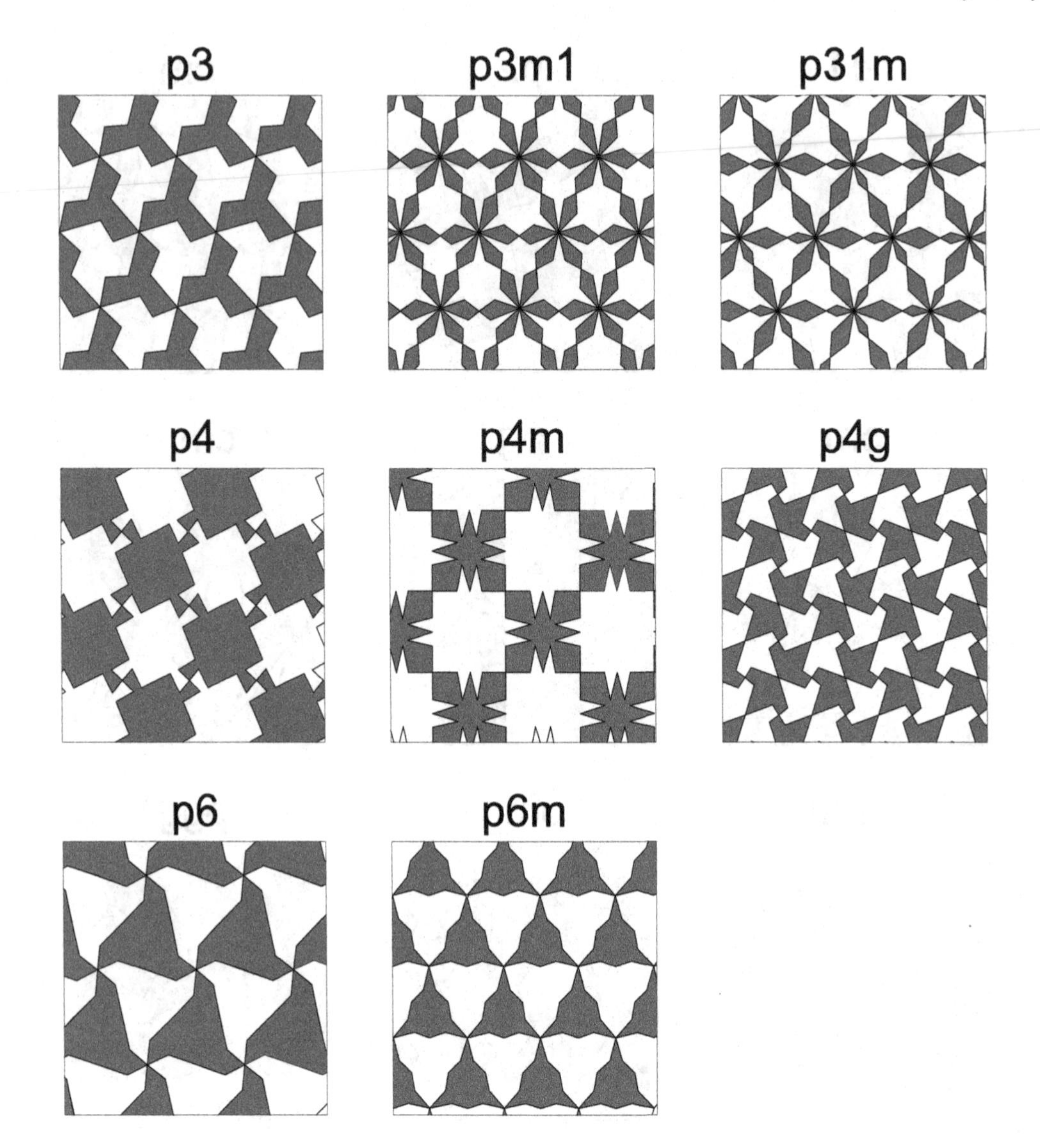
p3
p3m1
p31m
p4
p4m
p4g
p6
p6m

References

1. Mackiw G (1985) Applications of abstract algebra. Wiley, Hoboken
2. Schwartzenberger R (1974) The 17 plane symmetry groups. Math Gazette 58:123–131
3. Schattschneider D (1978) Tiling the plane with congruent pentagons. Math Mag 51:29–44
4. Schattschneider D (1978), The plane symmetry groups: their recognition and notation. Am Math Monthly 85(6):439–450

Correction to: Identification Numbers and Modular Arithmetic

Correction to:
Chapter 1 in: D.R. Finston and P.J. Morandi, *Abstract Algebra: Structure and Application*, Springer Undergraduate Texts in Mathematics and Technology, https://doi.org/10.1007/978-3-319-04498-9_1

In the original version of Chapter 1, the figure was wrong which has now been replaced, at the end of the chapter opening page.

The updated version of this chapter can be found at
https://doi.org/10.1007/978-3-319-04498-9_1

D.R. Finston and P.J. Morandi, *Abstract Algebra: Structure and Application*,
Springer Undergraduate Texts in Mathematics and Technology, DOI 10.1007/978-3-319-04498-9_11

List of Symbols

Symbol	Meaning	Page
$a \equiv b \mod n$	Congruence modulo n	5
$\mathbb{Z}$	Set of integers	6
$\overline{a}$	Equivalence class of a modulo n	6
$\mathbb{Z}_n$	Set of integers modulo n	6
$\|G\|$	Number of elements in the set G	8
$\gcd(a, b)$	Greatest common divisor of integers a and b	11
$u \cdot v$	Dot product of vectors	18
$\mathbb{R}$	Set of real numbers	18
$\mathbb{Z}_2^n$	Set of n-tuples with $\mathbb{Z}_2$ entries	23
10101010	String notation for an n-tuple	23
$\mathbf{0}$	Zero vector	23
$\mathrm{wt}(v)$	Weight of a word v	24
$D(v, w)$	Distance between words v and w	24
$\lfloor a \rfloor$	Floor function	25
(n, k, d)	Parameters of a code	35
$C + w$	Coset of a code C	36
$A \times B$	Cartesian product of A and B	41
$\cup$	Union	42
$\cap$	Intersection	42
$\det(A)$	Determinant of A	42
$M_n(F)$	Ring of $n \times n$ matrices over F	43
$\mathbb{C}$	Set of complex numbers	43
$\mathbb{Q}$	Set of rational numbers	48
F^n	Set of n-tuples over F	58
$F[x]$	Set of polynomials in x over F	58
$C^\perp$	Dual code	70
A^T	Transpose of the matrix A	70
$\deg(f)$	Degree of the polynomial f	74
$\gcd(f, g)$	Greatest common divisor of polynomials	75
aR	Principal ideal generated by a	77
(a)	Principal ideal generated by a	77

D.R. Finston and P.J. Morandi, *Abstract Algebra: Structure and Application*,
Springer Undergraduate Texts in Mathematics and Technology, DOI 10.1007/978-3-319-04498-9

Index

D.R. Finston and P.J. Morandi, *Abstract Algebra: Structure and Application*,
Springer Undergraduate Texts in Mathematics and Technology, DOI 10.1007/978-3-319-04498-9

Printed in the United States
By Bookmasters